SELECTED READINGS IN PHYSICS
General Editor: D. TER HAAR

MEN OF PHYSICS

KARL LARK-HOROVITZ

FRONTISPIECE: Photograph of oil painting of Karl Lark-Horovitz executed in 1954 by Mr. Raymond Breinin, then of Chicago, now in New York—commissioned by colleagues, friends, and former students of K. L-H.

MEN OF PHYSICS

KARL LARK-HOROVITZ

Pioneer in Solid State Physics

BY

V. A. JOHNSON

Professor of Physics, Purdue University

PERGAMON PRESS

OXFORD · LONDON · EDINBURGH · NEW YORK

TORONTO · SYDNEY · PARIS · BRAUNSCHWEIG

PERGAMON PRESS LTD.,
Headington Hill Hall, Oxford
4 & 5 Fitzroy Square, London W.1

PERGAMON PRESS (SCOTLAND) LTD.,
2 & 3 Teviot Place, Edinburgh 1

PERGAMON PRESS INC.,
Maxwell House, Fairview Park, Elmsford, New York 10523

PERGAMON OF CANADA LTD.,
207 Queen's Quay West, Toronto 1

PERGAMON PRESS (AUST.) PTY. LTD.,
19a Boundary Street, Rushcutters Bay, N.S.W. 2011, Australia

PERGAMON PRESS S.A.R.L.,
24 rue des Écoles, Paris 5e

VIEWEG & SOHN GMBH,
Burgplatz 1, Braunschweig

First edition 1969

Library of Congress Catalog Card No. 77–91464

Printed in Great Britain by A. Wheaton & Co., Exeter

08 006580 5 (flexicover)
08 006581 3 (hard cover)

Contents

Preface

UP TO December 1941, Karl Lark-Horovitz was known to his colleagues in the United States and abroad for important work in structure determination by X-ray and electron diffraction methods, for his vigorous pursuit of knowledge in the then-young field of experimental nuclear physics, and for his herculean efforts in creating a significant graduate programme and research effort at Purdue University during the preceding decade.

In January 1942 he was invited by H. M. James, who acted as a representative of the Radiation Laboratory (operated for the U.S. Government by the Massachusetts Institute of Technology), to participate in the war research effort by carrying out for the Radiation Laboratory certain investigations related to the improvement of radar techniques and devices. Lark-Horovitz agreed to do this by working on "crystals". Furthermore, he suggested that he and his associates concentrate, at least initially, on the element germanium by studying its physical properties and its use in devices, especially rectifiers. During the next three years Lark-Horovitz and the members of his Purdue group achieved the production of high-purity germanium, learned how to control the electrical behaviour of germanium specimens by the systematic addition of properly chosen impurities, interpreted the observed electrical and thermoelectric behaviour of germanium, and applied their knowledge to the improvement of devices used in radar.

The war-stimulated research carried out at Purdue University, along with related studies made at a number of other U.S. universities, several industrial laboratories, and certain laboratories in Great Britain, strongly stimulated the growth of the field that came to be called solid state physics. After 1945 this field expanded explosively, both in terms of numbers of physicists involved and in variety of problems investigated.

Upon the cessation of World War II hostilities, Lark-Horovitz chose to continue and extend his studies of germanium and other semiconducting materials. He initiated research on the effects of radiation upon the behaviour of semiconductors. His world-wide reputation as a solid state physicist was greatly enhanced by these radiation studies. Quite appropriately, the bulk of his publications reproduced in this volume deals with this topic.

It was my privilege to be a graduate student at Purdue during the 1930's and thus to participate in a small way in the building of the department. I was selected by Lark-Horovitz in 1942 to join the group studying the characteristics of germanium and other semiconductors. From then on until 1958 I was fortunate in having the opportunity to collaborate with him on a number of investigations in solid state physics. Much of the material in Part 1 is derived from my personal knowledge of the events described.

I have also had the privilege of drawing on the material gathered by four of my colleagues for an informal memorial meeting held in April 1958. I thank Professors H. Y. Fan, D. J. Tendam, I. Walerstein, and H. J. Yearian for their courtesy in allowing me to quote from their remarks on this occasion. I also thank Professor H. M. James for his permission to quote from a memorial statement that he prepared. Professor R. W. Lefler has been very helpful in providing material pertaining to Lark-Horovitz's service on the American Association for the Advancement of Science Cooperative Committee. I also thank these gentlemen for their suggestions during the course of preparation of the manuscript, and I also acknowledge the encouragement given by Professor R. W. King, the present Head of the Purdue Physics Department, and appreciate receiving his permission to draw material from the files of the Department.

Finally, Mrs. Betty Lark-Horovitz, who has read the entire manuscript, offered a number of cogent suggestions for its improvement. She has also supplied some details of the early life of her late husband.

I also express my appreciation to Mrs. S. Louisa Spencer for her assistance in the preparation of the manuscript.

West Lafayette, Indiana VIVIAN A. JOHNSON

PART 1

THE MAN

I

Introduction

KARL LARK-HOROVITZ was a latter-day Renaissance man, a man of many parts, a man with tremendous impatience and unbounded enthusiasm in his eager search for knowledge. No field of intellectual or artistic endeavour failed to capture a share of his interest and attention. The breadth of his viewpoint and range of his education enabled him to achieve professional success which may be classified into four fields: (i) research in the wide variety of subfields of physics in which he personally participated, (ii) the creation of a highly regarded graduate programme in physics at Purdue University, a feat involving both the acquisition of facilities in the face of tremendous difficulties and the stimulation of young physicists to develop their own talents to the utmost, (iii) service to the educational system of the entire United States by his active participation in the creation, development, and fruition of the AAAS† Cooperative Committee on the Teaching of Science and Mathematics, and, finally, (iv) his personal efforts and achievements in the renascence and unfolding of the field now known as solid state physics.

He was born Karl Horovitz in Vienna, Austria, on July 20, 1892, the son of Moritz and Adele Horovitz. His father, a noted dermatologist, pursued scholarly interests in fields as divergent as botany and the works of the Roman poet Horace. With such a family background, it is not surprising that the young Karl attended the humanistic high school rather than the engineering one. He entered the University of Vienna in 1911 and pursued studies in chemistry, physiology, physics, and pre-Socratic

† American Association for the Advancement of Science.

philosophy. His studies were interrupted for four years by service as an officer in the Signal Corps of the Austrian Army and recuperation from wounds received in service. He married on July 26, 1916. His wife, Betty, is still living.

In 1919 Horovitz returned to the University of Vienna and received his Ph.D. degree in physics later that year. The development of his productive professional career will be taken up in the ensuing chapters.

One of his fields of scientific interest was the physics and chemistry of glass. He also had an avocational interest in glass and other ceramics, acquired a personal collection of fine quality, and responded most willingly to the many invitations to discourse on the history and technique of glass-making at a "popular" level. He had a considerable knowledge of music and played the violin and viola. In his early years at Purdue University he organized a string quartet which played regularly. He was a member of one chamber music group or another from that time until he was forced out of participation by the demands of his war-time activities in research and education. Those fortunate enough to meet him readily became aware of his ability as a conversationalist and his solid acquaintance with literature, art, history, and politics.

Most impressive of all his attributes was his complete devotion to science. Professor H. Y. Fan, colleague and friend of Karl Lark-Horovitz, spoke† about this characteristic as follows:

"We have all seen his bad humour at times. Perhaps he was loaded down with uninteresting work, such as problems of administration. When you came into his office he would be frowning, but as soon as you began to explain to him the research problem that puzzled you at the moment, or to describe some interesting result just obtained, you immediately saw the sign of interest reflected in his eyes. He never became impatient when a problem of real scientific interest was discussed. This devotion to science enabled him to be a strong administrator, because he felt compelled in the interest of science to attend personally to various

† Quotation from words spoken in April 1958 at an informal memorial meeting.

phases of departmental affairs, and to be at the same time an active research worker in a very true sense of the word. You may have seen him talking on the telephone to people in the administration building, or long-distance to some one in Washington, talking about all sorts of things, but the moment he put down the phone he would pick up the trend and continue the discussion of the research problem as if there had been no interruption. This could come about only because he was really interested in the physics he was discussing with you."

The family name was changed in 1926 from Horovitz to Lark-Horovitz. In doing this, Karl Lark-Horovitz linked his wife's name as an artist-print-maker to his own, a change that was the result of a definite wish on his part. Later, he wanted his children to carry on the name of Lark only. Karl and Betty Lark-Horovitz had two children, a daughter, Caroline Betty, and a son, Karl Gordon, born in the United States in 1929 and 1930, respectively. Karl Gordon, like his father, is a scientist. He and his wife, Cynthia Thompson Lark, are engaged in research in the field of molecular biology, in which they have published widely.

Karl Lark-Horovitz died, of a heart attack, on April 14, 1958, while at work in his office in the Physics Building of Purdue University.

II

The Early Years (1914–1928)

KARL HOROVITZ's first publication† appeared in 1914, during his student years at Vienna. It dealt with the historical development of the physical concepts of relativity, a topic consistent with his lifelong interest in the history of science. Two more papers were prepared before his studies were interrupted by World War I. These are indicative of his bent to explore in newly opened fields. In this case Fritz Paneth and he used radioactive materials (ThB, ThC, Po, and Ra) contained in solutions of various acids and bases to study adsorption by a number of different salts.

After receiving his Ph.D. in 1919 Dr. Horovitz stayed on at the University of Vienna until 1925 in a position combining research and teaching. His first project combined his interests in physics and physiology and led to a series of papers on the physics of image formation by the human eye. In particular, he described the influence of physical phenomena on visual acuity and the apparent image size (Selection 1 of Part 2).

Horovitz also carried out a study resembling his earlier work with Paneth in that radioactive materials were used in the investigation of a physical process. He used ThB and ThC to study the structure and surface characteristics of platinum crystals. His final work at Vienna was concerned with electrochemistry, primarily the electrode functions and other electromotive properties of glasses.

"It was in 1925 that Karl Horovitz was awarded an International Research Fellowship by the Rockefeller Board.‡ This was the

† A complete list of publications is given at the end of Part I.

‡ Quotation from words spoken by Prof. I. Walerstein of Purdue University in April 1958. Walerstein was at Toronto during Lark-Horovitz's stay there.

second of these awards given. This brought him to the American continent, and at first to Toronto, where he spent a year on work which he had already begun in Vienna, namely the investigation of crystal forms of certain materials by X-rays. These were materials not ordinarily found in simple crystal form, but which could be deposited on a very cold surface and then subjected to powder methods of X-ray analysis. The reason for going to Toronto was that he wanted to investigate crystal properties at the low temperatures required for condensation and also to see what does happen at extremely low temperatures such that the reduction in lattice vibrations would sharpen the lines. At that time Toronto was one of the few centers in the world where low temperatures down to liquid hydrogen and liquid helium could be obtained.

"This research fellowship did not confine him to one place. He was permitted to go to other locations. He went to the University of Chicago, then to the medical research laboratory of the Rockefeller Institute, and from there to Stanford University. The lines of research followed at those institutions were not always the same. The low temperature and structure work were pretty much finished up at the University of Chicago. He turned his attention to the measurement of the permeability of cell walls to various types of ions [Selection 2 of Part 2]. At the Rockefeller Institute he worked on problems in the line of biophysics with K. Landsteiner and was stimulated by the presence of A. Carrel and other distinguished scientists."

Lark-Horovitz presented the first of his many contributed papers to the American Physical Society in Philadelphia at the Christmas 1926 meeting. He described a technique for depositing metal films on glass surfaces in high vacuum and then observing the X-ray diffraction pattern of the film while it is still in vacuum and thus free of surface contamination. Four months later he talked about measuring the surface areas of various charcoal adsorbers through measurement of the amount of sodium oleate adsorbed from solution, with it known that the sodium oleate forms a monomolecular layer.

III

Building a Department (1928–1941)

"IN THE spring of 1928 Lark-Horovitz was invited by R. B. Moore, Dean of the School of Science, to deliver a series of lectures at Purdue University.† Moore had been appointed Dean just about a year earlier; he was also Head of the Chemistry Department. He was a man of scholarly ability, forward looking in research, and he wanted to see that a good Physics Department should also exist on the campus. He was able to persuade Lark-Horovitz to return in the fall of 1929 to assume a permanent position at Purdue.

"The type of intellectual desert into which he came at that time may be a little difficult for some of you to imagine. The department consisted of five professors, one of whom had a Ph.D., and six graduate assistants. The rule on the campus a year or so before his coming here was that nobody was allowed to enter the buildings at night, nor permitted to have a key to the building. I don't know what they would have wanted a key for, because the library facilities of the Physics Department consisted of one corner of a very small room that had a few shelves of text books for general physics. People were told that if they wanted to do research work they would have to do it at their own expense and on their own time. The teaching load for a graduate assistant was fifteen hours per week or more. The equipment available for research was essentially nil. The number of courses offered for graduate work was a quantity which could be described as epsilon. I think that would have been sufficient to dissuade most persons from accept-

† Quotation from I. Walerstein, April 1958.

ing the position, and it is characteristic of Dr. Lark-Horovitz that he accepted this challenge and responsibility."

How did Lark-Horovitz react to his first experience with the Purdue Physics Department? The following is quoted from a report that he wrote to Dean Moore in February 1929, a year after his initial arrival:

"The physics department of Purdue University at the time of my arrival existed mainly as a service department for the different schools. The training of the professorial staff and the assistants had therefore been only routine teaching, and, with the exception of Dr. R. B. Abbott, nobody was doing any research work or was interested in such work. The equipment was only suited for the different courses in undergraduate physics.

"The purpose of the special lectures I was to give, as outlined by Dean Moore, was the following: To stimulate interest in modern physics; to give a survey, for the staff of the different departments of the School of Science, of the ideas and conceptions prevalent in modern physics; and to discuss on a general basis the relations between modern industry and research in physics.

"Three courses were offered:

(a) Electron Theory of Matter, including a mathematical introduction and the theory of relativity;

(b) Atomic Structure, including a general survey of the experimental work in radioactivity and chemical physics;

(c) Eight popular lectures with experiments, lantern slides and demonstrations, on Science in Industry.

"After the first lecture we were informed and soon found out for ourselves that the mathematical training, with the exception of a few mathematicians and chemists, was not equivalent to the requirements of the courses. We tried, however, to maintain as high a standard as possible, limiting the mathematical discussions only to the most necessary implements."

How, then, did Lark-Horovitz take the first steps to change the department from the dismal state just described into the respected, productive status reached by the end of the thirties? He had full understanding and support, not only from Dean Moore (who died in 1931), but especially from the University President, Dr. E. C. Elliott, who met Lark-Horovitz's requests within budget limitations. Among the steps taken, starting as early as 1929, were:

(a) The formation of a physics library adequate as a research tool. Subscriptions were entered to periodicals such as the *Physical Review*, *Annalen der Physik*, *Zeitschrift für Physik*, *JOSA* and *RSI*, *Proceedings of the Royal Society* (*of London*), etc., and a number of back runs were also acquired. A number of German, British, and American books at the advanced level were purchased, including volumes of the *Handbuch der Physik* and the *Handbuch der Experimental-Physik*. This nucleus of a library was housed along one wall of a classroom that was also used as an office and a seminar room.

(b) Setting aside space for an instrument shop, which was equipped with a lathe, drill press, milling machine plus some bench tools and staffed by one full-time machinist and one half-time man. In this way research equipment too expensive to purchase could be built in the department through the combined efforts of the shop staff and the graduate student working on the research project. A glass shop was also set up; initially it required the half-time services of a glass-blower.

(c) The use of the good services of Dr. G. S. Meikle of the Purdue Research Foundation plus Lark-Horovitz's own contacts to beg from industrial laboratories major pieces of equipment that were obsolescent for industrial purposes. He also persuaded Purdue's great industrialist friend and alumnus, David E. Ross, to provide about $1100 for the materials and labour to construct an electromagnet in the Physics Department shop (heavy-duty machining and forging were done elsewhere).

(d) The addition of three new professors with Ph.D. degrees to the Physics Department staff. These were young and able men with plenty of enthusiasm for tackling the formidable job of building a going research community where there previously had been none. Recognizing that the basic strength of a university department is determined by the number and quality of its graduate students, Lark-Horovitz obtained the funds necessary to hire 15 new graduate assistants during his first three years at Purdue. Through the contacts he had made while travelling as an International Fellow he was able to recruit young people who were well prepared at good schools. He also obtained industry-sponsored funds for bringing to the department five research fellows during this initial period. President Elliott assisted by authorizing the reduction in teaching loads from 15 to 9 hours per week.

(e) Persuading the University administration to bring two able applied mathematicians from Europe to Purdue. The courses offered by Professors C. Lanczos and W. E. O. Maier were of great value to the graduate students of physics during that period.

(f) Bringing to the campus a number of well-known physicists. These visitors would give one or several lectures on their own research specialities and freely consult with Purdue students and staff members during stays ranging from a day or two to a few weeks. Among those coming on such a basis before mid-1935 were Hans Bethe, J. C. Slater, Otto Stern, K. F. Herzfeld, Guido Beck, and Richard Courant. In February 1935 Lothar W. Nordheim arrived as a Visiting Professor. He added to the staff a person expert in modern theoretical physics, including quantum mechanics and the theory of metals.

Professor H. J. Yearian, who was the first Ph.D. candidate to complete work under the direction of Lark-Horovitz, in 1934, has described these early years of Lark-Horovitz at Purdue from the viewpoint of an experimental physicist:

"Even with a shop the beginning projects had to be chosen with great care. Whatever one chose, it required considerable ingenuity on the part of both the student and the director to get something started. His broad background of interests in science did enable Lark-Horovitz to make a wise selection of problems, quite a number of which were in fields in which he had had previous experience. They were mainly what might be called the physics of surfaces—studies of surface tension, interfacial tensions, mostly in organic acids; continuation of earlier work on ion exchange between glasses and electrolytes; the glass electrode and the dependence of its electrode reaction on the composition of the glass, the acidity of the solution, and so on. This work also led to further studies on adsorption, and in some of these radioactive tracers were used again. The study of orientation of molecules and ions on surfaces led to the first sponsored research, which was a project on friction electricity supported by the Eastman Kodak Company. This was a problem of considerable interest to them in the preparation and manufacture of film.

"Lark-Horovitz' past interest in the chemical behavior of glass led to an interest in the physical properties of glasses. There was a project running through most of the 30's on these properties, chiefly the viscosity and the electrical resistivity as functions of temperature. This was the second sponsored research, supported by Corning Glass Company.

"Most of the problems just mentioned were started in 1928. Then, when Dr. Walerstein came in 1929, the spectroscopy laboratory was founded. The E-1 Hilger spectrograph was the first large capital item at $2100 (a fifth of the capital budget). This year also marked the first outright grant to the department in the form of the Ross magnet, which was used by Walerstein in investigating the Zeeman effect. There were some associated researches in spectroscopy, carried out under Lark-Horovitz' direction, concerning Raman spectra. These could be done with rather modest equipment and so it was a good line to take up.

"Lark-Horovitz' tremendous interest in music and musical instruments led to another line of research. He encouraged Prof. Abbott to run acoustical analyses on instruments, chiefly violins, trying to find from analyses of the acoustical frequency spectra what distinguished a good violin from a poor one. This was rather difficult, but the work was carried on for a number of years.

"Another interest led to X-ray and electron diffraction work, which was the largest single line of research during the 30's. It was expedient to start this work in 1928, because Lark-Horovitz had in his past researches accumulated some X-ray equipment which he brought with him. He had two continuously pumped X-ray tubes and a vacuum camera, so that two students could start research immediately. F. E. Washer was concerned with the low temperature crystal structure of carbon disulfide, and I. G. Geib with testing some of the predictions of the dynamic theory of reflection of X-rays from crystals.

"About 1930 more students came into the X-ray laboratory and construction of equipment could be carried on so that eventually many lines of research were followed. There was a problem concerned with the rotation of the plane of polarization of X-rays when scattered from ferromagnetic material. There was a long series of structure investigations of fibrous semicrystalline and preferentially oriented materials, including woods. Other research dealt with the structure of truly amorphous materials. Nearly all of the alkali halides were investigated in the molten state to correlate the size of the ions and the type of packing in the liquid with the structures found in the solid. Sulfur and vitreous selenium were also investigated.

"Along with the building of the X-ray research facilities, a good deal of equipment was built and set up as a student laboratory. The student laboratory course was started in 1932. It quickly grew to a full-fledged course in X-rays, one of the few in the country at that time.

"Associated with the development of the X-ray laboratory was the electron diffraction work, which began with Yearian's research in 1929. That was only two years after the discovery of electron diffraction by Davisson and Germer and by G. P. Thomson. The research in this general field covered multitudinous things, but the basic concern at Purdue was with the intensity of electron diffraction.

"This bare outline might give the impression that these things came easily. Actually they came with great difficulty in view of the lack of equipment. Originally there was nothing with which to work. Everything had to be constructed. Very, very little could be purchased. This situation drew very strongly upon the ingenuity of the individual students. As an example, in 1929 practically all the vacuum systems in use were glass systems with mercury and glass diffusion pumps. We had a number in the department, but we had no glass blower. It was about two years before we had even a part time glass blower."

Even while devoting so much time and effort to building up a research group and its equipment, Lark-Horovitz turned out a considerable amount of research. After only four months at Purdue, he and G. W. Sherman (who taught pyrometry in the Department of Physics) presented to the American Physical Society a paper describing the use of a photoelectric cell to control, by feedback, the operation of any system whose state can be registered by mirror reading instruments.

X-ray studies of the crystal structure of metals, started during the International Fellowship phase, were extended to the study of mercury. Lark-Horovitz found that the crystal structure of this metal is the same at liquid air temperature and at the temperature of a dry ice–alcohol mixture, thus establishing the non-existence of a previously suspected structure change at − 80°C.

Lark-Horovitz worked with a Corning Glass Research Fellow, J. E. Ferguson, on the problem that had interested him while he was at Stanford, i.e. the electromotive properties of dielectrics. They made measurements of the electrode potential of quartz glass vs. a hydrogen electrode in acid solutions, vs. a Na electrode in NaOH and NaCl solutions, and vs. a Ag electrode in silver salt solutions. They also studied the electrode potential of paraffin vs. other electrodes in suitable solutions. These results are presented in Selection 3 of Part 2.

A little later Lark-Horovitz started work with a second Corning Glass Research Fellow, C. L. Babcock, that led to one of the early Ph.D. degrees granted by the Purdue Physics Department. This problem consisted of measuring the electrical conductivity and viscosity of several different types of glasses while in the molten state. These quantities were measured as functions of temperature, and an attempt was made to ascertain the fundamental principles determining the conductivity and viscosity properties.

Ph.D. level research was carried out in X-ray crystallography by several graduate students working under Lark-Horovitz. Among these investigations were a study of the dynamic reflection of

X-rays from ZnS (I. G. Geib) and a study of the crystallization of polymorphous ZnS from the vapour phase *in vacuo* as it is deposited on various types of surfaces held at 20°C (S. E. Madigan). The Faraday effect, at a Ni surface, was studied in the X-ray region (H. T. Clark).

A general problem to which Lark-Horovitz gave his attention over a span of better than twenty years was the X-ray determination of the structure retained by materials as they pass from the solid into the liquid phase. In the mid-thirties his student E. P. Miller determined the diffraction ring patterns for many liquids (glycerin, different fractions of paraffin oil, commercial mineral oils, castor oil, and molten alkali halides—LiCl, LiBr, NaCl, KCl). In some cases the results yielded, through an ingenious application of Fourier analysis, coordination numbers, which were found to be the same in solid and liquid for nearest neighbours; however, the distribution of second nearest neighbours was found to be disturbed in the liquid. Additional studies showed that vitreous solid selenium and liquid selenium yield X-ray diffraction patterns corresponding to the same distances for nearest neighbours as well as equal numbers of atoms at these distances. These investigations are represented by Selection 4 of Part 2, a publication of Lark-Horovitz and Miller on the structure of liquid argon. A few years later Lark-Horovitz and G. C. Danielson made a similar determination of the structure of liquid nitromethane. A related structure study carried out under the direction of Lark-Horovitz was C. M. Parshall's investigation of latex, in sheet form, in the unstretched state and with elongations of 100%, 200%, and 250%.

An especially interesting study was carried out by Lark-Horovitz and W. I. Caldwell. As has been mentioned, Lark-Horovitz possessed a life-long devotion to music and was himself an accomplished performer on the violin and viola. Like many another, he wondered about the factors responsible for the high quality of performance of violins fabricated by the master craftsmen Stradivarius and Amati. Lark-Horovitz decided to use X-ray procedures to ascertain the wood structures of violins and thus see if certain characteristics were common to all high-quality violins. Twenty-four violins were investigated by using techniques designed to observe oriented fibre patterns. This group included rare and costly instruments made available by the Lyon and Healy Company in Chicago. Definite correlations were established between the types of wood used for the front and back of a violin and its tonal quality. These results are presented in Selection 5 of Part 2.

As mentioned above, Lark-Horovitz in 1929 encouraged H. J. Yearian to build an electron diffraction outfit in the Physics Department shop and to initiate research with it. Yearian put most of his effort into measuring and interpreting the electron diffraction patterns formed by films of various metals and oxides, especially ZnO, and for this work obtained the first Ph.D. granted within the Purdue Physics Department. Instrumentation in this field from 1934 to 1937 was greatly assisted by the efforts of another graduate student, J. D. Howe, who had a positive genius for the design and construction of research equipment. The gist of this electron diffraction work is contained in Selection 6 of Part 2.

A prime interest of Lark-Horovitz was the stimulation of young scientists. He was adept at spotting talent, and he never stinted of himself in encouraging and instructing those showing promise. One such student was Edward Mills Purcell, who was destined to receive the 1952 Nobel prize in physics, jointly with Felix Bloch, "for their development of new methods for nuclear magnetic precision measurements and discoveries in connection therewith."

Purcell entered Purdue as a freshman in 1929 and graduated with a B.S. in Electrical Engineering in 1933. "His interest had already turned to physics, and through the kindness of Professor K. Lark-Horovitz he was enabled, while an undergraduate, to take part in experimental research in electron diffraction."† This research led to the first publication bearing Purcell's name, as a

† Quoted from p. 232 of *Nobel Lectures—Physics, 1942–1962*, Elsevier Publishing Co., Amsterdam, 1964.

co-author with Lark-Horovitz and Yearian, a contributed paper presented at the Cincinnati meeting of the American Physical Society in November 1933. This abstract follows:

"Electron diffraction from vacuum-sublimated layers. *Phys. Rev.* **45,** 123 (1934). The material for investigation is condensed in a high vacuum onto a volatile substance (camphor, naphthalene, etc.) held at liquid air temperature. By letting the volatile support evaporate after the desired thickness is reached it is possible to obtain thin, free films of the condensed material of varying thickness without using any dissolving agent as it is necessary in other methods. In this way films of zinc and zinc sulfide have been obtained. The electron diffraction pattern of zinc agrees with the X-ray diffraction pattern and shows no irregularity in the intensity distribution as compared with the corresponding X-ray pattern except for the first two lines. Zinc sulfide in thin layers forms colloidal particles; in thicker layers it is crystalline and the position of the lines agrees with the X-ray pattern. The intensities, however, are different in a similar way as reported previously in the case of ZnO. Copper, when deposited in this way, forms films of either pure cuprous oxide (Cu_2O) or a mixture of copper and cuprous oxide. The intensity distribution of the Cu_2O pattern agrees with the intensity distribution of the corresponding X-ray pattern."

The work done by Purcell on the technique of forming extremely thin films was continued and extended by J. D. Howe. The final results were published in the paper listed as item 22 in the Bibliography at the end of Part I.

A few years later Lark-Horovitz had a hand in the development of the young Julian Schwinger, 1965 Nobel Laureate. Particularly, Lark-Horovitz provided for Schwinger a university environment during 1940–3 and protected Schwinger from sundry encroachments on his time, thus giving the young man ample opportunity for professional growth.

Lark-Horovitz and Yearian found that the electron diffraction patterns of zinc oxide show intensities markedly deviating from the ones calculated under the assumption of a spherically symmetrical electron distribution in the Zn atom. Possible explanations were suggested on the basis of a shift of the electron cloud with respect to the nucleus, a distortion of the crystal lattice, a distortion of the electron cloud, and particularly of the valence electrons, or effects of dynamical reflection of the electron waves not taken into account in the kinematic theory. For a final decision among these possibilities, X-ray diffraction patterns from a flat sample of ZnO were obtained and analysed by Lark-Horovitz and C. H. Ehrhardt. They reached the conclusion that the observed intensity anomalies in the electron diffraction patterns were due to distortion of the electron charge distribution and its consequent effect on the electron "form factor".

The year 1935 marked the beginning of nuclear physics at Purdue. This start is described in the words of D. J. Tendam as spoken in April 1958:

"Plans were started then for two machines, a Van de Graaff electrostatic generator and a cyclotron. Lark-Horovitz' interest in the Van de Graaff was connected with his previous work in electron and X-ray diffraction. He really wanted the Van de Graaff as a high voltage source for a neutron generator, in which neutrons would be produced by a deuteron beam incident on a heavy ice target. It was his hope that these neutrons could be used for neutron diffraction work. The machine was completed in early 1937 and ran with about 250 microamperes of deuterons at about 250 KV. Some later adjustments brought this up to about 500 KV, but, of course, all the attempts at neutron diffraction failed. It has only been with the intense neutron beams available in reactors today that neutron diffraction work can be carried out. This electrostatic generator was a good example of how equipment was built with practically nothing. The total cost was $825.62 plus the hard labor of two graduate students, R. E. Schreiber and W. A. Miller.

"In the case of the cyclotron it was quite clear from the beginning that there was going to be very little money available beyond the usual departmental budget, so the primary objective was to build a compact machine at the smallest possible cost. Lark-

Horovitz wanted this machine primarily for nuclear physics, but he also knew that one could do something else with a cyclotron. One could produce radioisotopes for tracer experiments. This gave him the chance to point out to the people who provided funds, that this machine was going to do a lot of good for a lot of departments on campus. Tracer experiments can be done in lots of fields. In this way, perhaps, he got a little more money than he would have otherwise.

"The Carnegie–Illinois Steel plant in Gary furnished the magnet steel for $1598, the cost of the actual labor and shipping costs. Professor Yearian and two graduate students, J. D. Howe and S. L. Hluchan, personally moved the steel from the railroad siding into the cyclotron space of the Physics Building in order to save $150. The magnet coils, still in use, were wound in the Physics Department Shops. The testing of the magnet and the field measurements were done by Professor I. Walerstein and J. D. Howe. They then went on with the construction of the mechanical parts and the vacuum system. Professors W. J. Henderson and L. D. P. King and Dr. J. R. Risser constructed the oscillator and the radio-frequency system.

"So, the construction was largely a matter of hard labor with very little cash outlay. The total cost of the cyclotron was $9030.90, not including the labor of Purdue Physics Department staff. The construction was completed in August 1938, and before the end of that year there was a beam of about 3 microamperes of 8 MeV deuterons."

In the spring of 1939 Lark-Horovitz, working with J. R. Risser and R. N. Smith, published the results of their experiments in which indium was bombarded with 16 MeV alpha-particles from the Purdue cyclotron at $0 \cdot 03\ \mu A$ for 30–90 min. Their work showed that the activation of the In^{115} leads to a capture process with the formation of radioactive Sb^{118}, which goes back to radioactive In^{115} with a lifetime of $3 \cdot 6$ min. Lark-Horovitz and his colleagues started a systematic search for $(\alpha, 2n)$ reactions. These experiments were interrupted by World War II, to be resumed later by Bradt and Tendam [*Phys. Rev.* **72,** 1117 (1947)].

As might be expected, Lark-Horovitz was investigating the fission process in the days right after its discovery and shortly before all work in this field "went underground for the duration". Selection 7 of Part 2 presents his most important work in this field.

The development of nuclear physics at Purdue was bolstered by frequent lectures presented by visiting physicists such as W. Pauli, J. R. Oppenheimer, E. Segré, R. H. Fowler, and M. Schein. In the fall of 1941 K. W. Meissner joined the Purdue faculty. He established a research programme in spectroscopy and carried on distinguished research in that field until his death in 1959.

As was indicated in the discussion of his early career, Lark-Horovitz had a deeply rooted interest in the application of physical techniques to the solution of biological problems and had worked on several facets of this area. He had also used naturally radioactive materials as process tracers in some of his earliest research and could rightfully be considered a pioneer in this field. As soon as the Purdue cyclotron came into operation, he started experiments in which artificially radioactive isotopes were used as tracers. He carried out these studies with Herta Leng, an AAUW† International Research Fellow, and Tendam. During 1939–41 studies were made on the uptake, distribution, and excretion of sodium and potassium in the human body, on the distribution of sodium and potassium in the human blood cell, and on the testing of enteric coatings for medications. Selection 8 of Part 2 presents the results of the research on enteric coatings. One aspect of the tracer programme was the investigation of the permeability of red blood cells in pathological cases. It was found that the presence of anaemia or syphilis, for example, alters the distribution of sodium and potassium between red cells and serum. Additional study attributed this change to disturbance of the electrolytic equilibrium in blood.

With the entry of the United States into World War II in December 1941, the research programme and other operations of the Purdue Physics Department were markedly changed. As

† American Association of University Women.

far as Lark-Horovitz was concerned personally, his long-time interest in the teaching of physics was stimulated by the nation-wide concern with the training of scientists, engineers, and technicians to meet the demands of a nation at war. The demands of advanced instrumentation for the war effort led to the involvement of many members of the Purdue Physics Department in the research programme of Section 14—Crystals—of NDRC (National Defense Research Corporation), operating out of the M.I.T. Radiation Laboratory. From this research as a base there developed the large solid state programme for which Purdue University is so well known. And it is for his work in leading this solid state group that Lark-Horovitz became best known on the national and international scenes.

IV

The Educator

As HAS been mentioned in Chapter III, one of the first acts performed at Purdue by Lark-Horovitz was the reshaping, updating, and expansion of the graduate level courses of the department. He firmly believed that graduate students should have the opportunity to hear current developments in physics described by persons having different viewpoints and approaches and coming from different institutions of learning. Since it was not practical, for considerations of both time and expense, to have graduate students engage in extensive travel to other universities, Lark-Horovitz made the presence of visiting scientists an "institution" in the Purdue Physics Department. These visitors came from all parts of the United States and from abroad. Sometimes a man gave a single lecture, sometimes several lectures, and, on occasion, a series extending over a summer session or a full semester. A list of these speakers includes names such as Oppenheimer, Pauli, Heisenberg, Segré, Lawrence, Rabi, and Teller.

Attention was also given to the improvement of the undergraduate courses offered. He focused his efforts on the demonstration lectures which formed the cores of these courses. He found able and enthusiastic professors to undertake the task of giving lectures to the students of engineering, physics, and chemistry. He, furthermore, supported the work of these professors by obtaining appropriate reductions in teaching loads and by acquiring adequate demonstration materials and equipment.

Lark-Horovitz himself for several years delivered the lectures in the physics course primarily intended for students of the biological sciences and pharmacy. He put into this job the same degree of enthusiasm that he did in his research activities.

Although his direct concern was primarily with the demonstration lectures, he fully supported other activities aimed at course improvement, e.g. studies of the effectiveness of testing procedures, textbook comparison studies, investigation of class sectioning procedures, and evaluation of various teaching techniques such as individual work in the laboratory in contrast with lecture demonstration laboratory work.

In April 1936 Lark-Horovitz gave an invited paper to the Society for the Promotion of Engineering Education, meeting in Chicago, on the historical viewpoint in teaching physics. He was a firm believer in the high value of a historical approach to the teaching of physics, especially because he felt that the student should be made familiar with the full development of ideas and concepts, how they occurred, what stimulated them, and also the interaction of science with other fields of human endeavour. He was a devoted student of the interaction, throughout the course of human progress, of physical science and natural philosophy with politics, literature, the arts, and the development of social institutions.

During the mid-1930's Lark-Horovitz served for several years as a member of the University Committee on the Education of Women. In this capacity he played a major role, along with Professors C. Lanczos and J. K. Knipp, in the design of survey courses in mathematics and physics. These survey courses were planned for the common core of a special degree programme known as the Liberal Science Curriculum. Initially, in 1939, thirty women students were admitted to this programme. At the end of 1939–40, Lark-Horovitz reported to Purdue President Elliott as follows:

"The survey courses in the exact sciences have met with great success. The students have shown a great deal of enthusiasm for physics, and their efforts have been far beyond what has been experienced in the past even with senior students."

Effective in the fall of 1940, forty women were admitted each year to this curriculum. During the 1940–1 year survey courses in chemistry and biology were added and, for the first time, the three sciences and mathematics were taught simultaneously in the Liberal Science programme.

After World War II the Liberal Science programme was enlarged in student capacity, men students were admitted as well as women, and survey courses were offered in an increasing number of departments. After 1950 the separate character of this programme was terminated. The experience and enthusiasm gained in this educational experiment served as the basis for building a science programme for humanities students in the newly organized School of Science, Education, and Humanities. During the 1950's Lark-Horovitz enthusiastically participated in the lectures of the physics course for humanities students and gave such a lecture only a few hours before his death.

Lark-Horovitz early realized that it was practically impossible to strengthen the work in physics in the colleges and graduate schools without improving science education in the secondary schools and so, in the early 1930's, initiated efforts intended to provide better training for high school science teachers. In 1935 he was awarded a citation by the American Association of Physics Teachers for his important contributions in the field of physics teaching, not only at the college level, but also to physics teaching in the high schools and the lower levels.

In co-operation with staff members of the Purdue Division of Education, Lark-Horovitz organized, by 1938, a number of courses to satisfy the needs of high school science teachers in Indiana and neighbouring states. The development of these courses was the direct outcome of a study on the teaching of physics in Indiana high schools carried out by Lark-Horovitz as chairman of a committee appointed by the Indiana Chapter of the American Association of Physics Teachers. This study produced definite information about the needs of physics and science teachers in the secondary schools and thus pointed the way to relieving these shortcomings through the use of Purdue University facilities. Just prior to this time, Lark-Horovitz had been a leader in organizing the Indiana Chapter of AAPT. During the late 1930's he participated actively on a Purdue committee on science teachers' training formed from representatives of the Division of Education and School of Science. This effort led to the development of a curriculum in the natural sciences and mathematics which stressed the co-ordination between different sciences and the related mathematics. The work of Lark-Horovitz directed towards improvement of science education continued into the war years. In 1944 he reported to the President of the University:

"The work started during 1942–3 on a physics teachers' guide has been extended by R. W. Lefler. A guide covering the whole year's work for the high schools has been published and is in use now in the Indiana schools. In co-operation with other groups, the physical science curriculum containing both physics and chemistry has been brought up to date, and suggestions of Indiana science teachers have been incorporated. In co-operation with local school systems, a special co-ordinated curriculum from the fourth to the twelfth grade in the sciences has been worked out and a special workshop conducted to make teachers familiar with the project."

The educational activities of Lark-Horovitz expanded into the nationwide picture in December 1939 when he was appointed, at the annual national meeting of AAPT, to a committee charged with the study of the teaching of secondary school physics in the U.S. and of ways of bringing about a better co-ordination of the natural sciences and mathematics in the schools. This committee work led shortly to the beginning of Lark-Horovitz's long service with the group that came to be known as the AAAS Cooperative Committee. The aims and achievements of this organization and the role of Lark-Horovitz therein are well described in the following quotations from a talk given at a meeting of the AAAS Cooperative Committee on October 12, 1965 to mark the twenty-fifth anniversary of the group:†

† *A Brief History of the AAAS Cooperative Committee on the Teaching of Science and Mathematics*, by Bernard B. Watson, Research Analysis Corporation.

"The Cooperative Committee on Science Teaching, as it was called initially, was created by representatives of five scientific societies in April 1941. The idea for such a committee was conceived by Professor Karl Lark-Horovitz of Purdue University who presented the idea to several groups of interested individuals beginning in December 1940. At these meetings it was generally agreed that there was need for cooperation among the scientific societies on problems of science teaching. Accordingly, the Committee was formed to work specifically on educational problems which no single scientific society could solve by working alone, e.g. certification of science teachers for high schools.

"Two representatives from each of the five societies formed the original Committee. These societies were the AAPT, Union of Biological Societies, Mathematical Association of America, American Chemical Society, and the National Association for Research in Science Teaching.

"The initial financial needs of the Committee were met by grants totalling $3000 from the Carnegie Foundation for the Advancement of Teaching.

"The original Committee served for three years without a change of personnel, and in the spring of 1945 was reorganized as a committee of the AAAS. At about the time of the change in sponsorship several other societies were admitted to membership on the Committee and additional societies have been added over the years.

"One of the urgent problems that the Committee addressed in its formative years, which coincided with World War II, was the problem of the supply of technically trained manpower for the armed forces and industry. In 1942–43 it prepared and issued a report on 'High School Science and Mathematics in Relation to the Manpower Problem.'

"A second problem on which it worked during its early years, and which has been a continuing major concern of the Committee, was the problem of state certification requirements for high school teachers of science and mathematics and the related problem of the college preparation of such teachers. During the war the Committee issued a preliminary report on this subject and followed it immediately after the war with its widely distributed Report No. 4 on 'The Preparation of High School Science and Mathematics Teachers.'

"In 1946 the Committee joined in an effort with a committee of the National Science Teachers Association to report on science course content and teaching apparatus used in U.S. schools and colleges. The Cooperative Committee prepared the college material and participated in an advisory capacity in the preparation of the entire report. This report, whose preparation was paid for by the Scientific Apparatus Makers of America, was submitted through UNESCO to the ministers of education of the devastated countries of the United Nations and provided guidelines for the reconstruction of war-damaged instructional laboratories.

"Perhaps the biggest job the Committee ever undertook was its assistance to the President's Scientific Research Board in 1947. The President's Scientific Research Board, under the chairmanship of John R. Steelman, Assistant to the President, was asked to report to President Truman on a number of matters concerning the status of scientific research work in this country. The information presented in this report to the President was intended to help the National Science Foundation get under way, should the Foundation have been established by the time the report was completed. If no Foundation were established by July 1, 1947, the report was to serve as a guide for future executive orders concerned with federal support for scientific research and education. As it turned out, the NSF was not established until several years later, and the report served as powerful support for proponents of the Foundation.

"The Cooperative Committee was asked, specifically, to consider and report to the President's Scientific Research Board on the question: What is the effectiveness of science training to increase the quality as well as the supply of scientists for government, industry and university research, and for high school and college teaching? The Committee's report makes up the major part of Vol. 4 of the Steelman Report on *Manpower for Research.*

"The most ambitious undertaking of the Committee and the one that has had the most far reaching effects directly attributable to the activities of the Committee was the development of what was first called an 'Action Program to Meet the Shortage of Well Qualified Science and Mathematics Teachers.' The outline of the program was developed by a Joint Subcommittee during the spring and summer of 1954 and was approved by the Cooperative Committee in October 1954. The program was renamed the 'Science Teaching Emergency Program' and later, when it became apparent that the emergency was going to last for a long time, the 'Science Teaching Improvement Program.'

"The Program as developed by the Committee consisted of six major projects:

"I. To encourage departments of science and mathematics in colleges and universities to accept the training of secondary school teachers as a major responsibility.

"II. To increase the number of qualified teachers on an emergency basis.

"III. To interest high school students in preparing for teaching careers in science and mathematics.

"IV. To support higher salaries for teachers.

"V. To promote improved working conditions for and increased efficiency of secondary school teachers of science and mathematics.

"VI. To provide for the recognition of exceptionally able teachers.

"The Committee proposed a three-year budget for the program totaling over one-half million dollars. The Committee recognized that it could not itself administer a program of such magnitude and recommended that the AAAS assume responsibility for the program.

"Prior to the next meeting of the Committee in February 1955 the AAAS Board of Directors approved the program developed by the Committee, adopted it as an AAAS program, and directed Dr. Wolfle† to seek at least $25000 to initiate the program and to appoint a director for it as soon as possible. Dr. Wolfle did a lot better and obtained a grant of $300000 from the Carnegie Corporation to underwrite the program for the first three years. John Mayor was appointed director of the STIP and reported for work on September 12, 1955.

"What started out as a program of limited scope and duration has since become an enlarged and permanent part of AAAS activities with Mayor as AAAS Director of Education. There is no doubt that a major share of the credit for the direct involvement of AAAS in matters of education belongs to the Cooperative Committee. . . .

"In my opinion one of the most important contributions of the Cooperative Committee, particularly in its early years, and one which it is difficult or impossible to document, was in prodding the conscience of scientists to take an interest in the problems of science teaching. It is hard to imagine now, with millions of dollars being poured into science teaching projects by the NSF and by private foundations, and with the Zacharias‡ and other prominent research scientists involved in these efforts, that it is only within about the last fifteen years that this type of activity has become acceptable for research scientists. Prior to that time scientists actually put their scientific reputations in some jeopardy by showing too much of an interest in teaching and its problems. Lark-Horovitz and E. C. Stakman, a well-known plant physiologist from the University of Minnesota and an early member of the Committee, were taking a chance with their reputations as scientists in becoming too closely associated with efforts of this kind. The Cooperative Committee deserves a large measure of the credit for keeping alive an interest in science teaching problems during the lean years when this was not popular and for encouraging individual scientists and professional societies to

† Dr. Dael Wolfle, AAAS executive.

‡ This refers to Dr. Jerrold R. Zacharias of M.I.T. who served as Chairman of the Physical Science Study Committee. This group revised methods of teaching high school physics.

actively engage in efforts toward the improvement of science teaching. Many of the activities in which the scientific professional societies are engaged today were stimulated directly or indirectly by the work of the Cooperative Committee."

Not only was Lark-Horovitz an original member of the AAAS Cooperative Committee, but he served as chairman of the Committee from 1945 to 1950. It was during this period that the Committee made its contribution to the aforementioned Steelman Report on *Manpower for Research*. This work was not the only contribution of Lark-Horovitz to the AAAS. He was General Secretary from 1947 to 1949 and then a member of the AAAS Editorial Board from 1949 to the time of his death in 1958.

In his final years Lark-Horovitz was concerned with two educational problems: the presentation of science to the non-scientist and the presentation of science such as physics to specialists in other fields of science. His thinking in this field is presented in the following abstract of a paper that he presented on December 29, 1953 to a National Science Foundation Symposium on "Science in General Education":

"*The role of physics in general education.* There are two problems for physics in general education: physics for the layman, the business man, the industrialist, economists or legislators, and physics for scientists, such as biologists, chemists, engineers, geologists, and medical men.

"The first problem has been discussed many times and various solutions have been proposed; in the schools and colleges special courses emphasizing concepts, not formalism; for adult education, university extension programs and popular magazines. Special efforts have been made in recent years by science writers to bring science to the people.

"For the scientists there are a great many physics programs, teaching and training in special techniques. They contribute little to the general education of the scientist. We need: (1) Cooperative research and teaching programs, staffed from various departments. Thus we will learn to understand the problems of neighboring fields and how physics can contribute to their solution.

(2) Journals maintained by scientific associations such as the AAAS, which bring frequent reviews, up-to-date and competent, for the information of scientists in neighboring fields. (3) A strengthening of borderline fields, such as biophysics, medical physics, geophysics, and engineering physics. (4) Clarification of the social implications of science and the impact of physics on society."

In summary, Lark-Horovitz played an active role in the reorganization and the teaching of undergraduate courses from the time of his arrival at Purdue until his death. He integrated ideas from chemistry, biology, and philosophy into the program of the general physics courses. He organized his material to emphasize the growth of the science of physics as one aspect of the history of civilization and to show the effect of scientific discovery on the social and economic development of society. His activities as a member, and then as Chairman, of the AAAS Cooperative Committee were indicative of his strong interest in the teaching of science at all levels, from kindergarten to graduate school, in all places, whether Purdue University or any other school or college in Indiana, in another state of the United States, or even in another country.

V

Pioneer in Solid State Physics (1942–1958)

"SOLID State was the field of research of Dr. Lark-Horovitz in which he has perhaps made the most important contributions to physics, and the field in which he was recognized as one of the leading authorities.† Without exaggeration one might say he was one of the pioneers in the modern research on semiconductors, and it is not just we who are in the Purdue physics department who think so; the recognition is indeed international. I recollect the 1954 international conference on semiconductors in Amsterdam, which Dr. Lark-Horovitz could not attend on account of ill health. There was a unanimous resolution of the conference to send a telegram expressing regret that he could not be there.

"His work in solid state started at the beginning of the war, early in 1942, when the Radiation Laboratory expressed interest in crystal rectifiers and was looking for laboratories to assume subcontracts to investigate crystal rectifiers. Dr. Lark-Horovitz at that time, in discussion with the people at the Radiation Laboratory, chose the material germanium to work with. As you know, germanium has played an important role in the development of semiconductor physics. Dr. Lark-Horovitz in choosing this material showed great insight."

The early progress of this research was outlined later by Lark-Horovitz in a summary that he called *History of Germanium Development at Purdue:*

Jan. 1942 Dr. James submits a number of problems to be worked on outside of the Radiation Laboratory,

† Quotation from words spoken by Prof. H. Y. Fan in April 1958.

among them the problem of crystal detectors. Because of my experience in this field it was offered to the Radiation Laboratory that the Purdue group should engage in this type of work.

Jan.–Feb. 1942 — Visits to various installations and discussions to learn the present status of detector development. Literature study. It was found that essentially the silicon detector, in the form in which it was used by the English and the Americans, was developed by the Germans before the war. Sperry Gyroscope Laboratory had introduced at this time crude germanium as a detector. Literature studies on germanium detectors.

March 1942 — Organization of the Purdue group with the program to purify germanium. Whaley's experiments on purification of germanium using all methods known at this time. First experiments on melting under helium, hydrogen reduction, etc.

May 1942 — Meeting at M.I.T. Present: E. U. Condon, representing Westinghouse, F. Seitz, representing University of Pennsylvania, H. Q. North, representing General Electric, T. A. Becker, representing Bell Telephone, N. Rochester, representing Radiation Laboratory, K. Lark-Horovitz and R. G. Sachs, representing the Purdue group. At the beginning of the meeting, K. Lark-Horovitz announced the production of p- and n-type germanium by addition of either boron, aluminum, gallium, indium (this series was at the time completed) or arsenic, bismuth from the other series. The next day North approached K. Lark-Horovitz and asked for permission to work on germanium at General Electric.

Development of purification and production of larger ingots during May, June and July. First detector units produced in the summer of 1942.

August 1942 Visit of Rochester to Purdue and assignment to investigate "burn-out" in germanium crystals. Meeting at Columbia University. Lark-Horovitz announced for the first time that germanium and silicon are intrinsic semiconductors, as substantiated by findings at the University of Pennsylvania and also by findings in the literature, but not recognized before.

Sept. 1942 During burn-out experiments Benzer discovered that welded units with whiskers will still rectify. Lark-Horovitz pointed out that D.C. welding might be used for production of units. High-back-voltage characteristics observed in some material by Benzer. Continuation of electrical measurements by Lark-Horovitz and his group (E. P. Miller, I. Walerstein, later joined by A. E. Middleton, W. W. Scanlon); this is basic for applying the diode theory developed by H. A. Bethe, Radiation Laboratory. Investigation of gas absorption on the detecting properties of semiconductors, planned by Lark-Horovitz and assigned to Whaley. Purdue group divided into three essential units: (a) electrical measurements of Hall effect, resistivity, thermoelectric power under K. Lark-Horovitz, (b) purification and melting—R. M. Whaley, (c) burn-out and high-back-voltage rectifiers—S. Benzer, R.F. properties—H. J. Yearian and R. N. Smith, theory—first R. G. Sachs, then V. A. Johnson.

Spring 1943 High back voltage observed first in Whaley's high vacuum experiments. Continuation of these experiments by Benzer led to high-back-voltage diode.

Summer 1943 High back voltage observed up to 150 volts and reported at Radiation Laboratory meeting in October 1943.

October 1943 Conference with H. Q. North, General Electric,

pointing out the possibilities of future germanium development. Assignment of mass production to Bell Telephone Laboratories. Purdue has the duty to supervise development and to meet regularly every six months at the Bell Telephone Laboratories with a group from NDRC† and a group from the armed services.

Spring 1944 Purdue group (V. Johnson, K. Lark-Horovitz) succeeds in interpreting resistivity and thermoelectric behaviour of germanium semiconductors. R.F. testing methods introduced by R. N. Smith (Purdue) are accepted by all NDRC groups. Reports on capacity measurements by R. N. Smith, high frequency measurements by Yearian, measurements of the static characteristics and determination of the rectification coefficient. General attack on the problem of how to predict R.F. characteristics from D.C. properties.

Fall 1945 At the end of the war the Purdue group had (a) shown electrical properties to be predictable from impurity content, (b) predicted resistivity and thermoelectric power in the range of temperature available at this time (down to liquid air temperature) from the number of electrons given by Hall effect measurement, (c) determined the mobility ratio for holes and electrons. First infrared measurements by K. Lark-Horovitz and K. W. Meissner yielded the dielectric constant for Si $\approx$ 13, for Ge $\approx$ 16–17.

High-back-voltage rectifiers were perfected and the present-type cartridge introduced by R. N. Smith. Methods of melting and production of high-purity ingots brought to high perfection by

† National Defense Research Corporation, which was set up to write contracts on behalf of the U.S. Government with university and industrial laboratories.

R. M. Whaley. The group decides to abandon development of detectors and the practical applications and to concentrate primarily on the basic investigation of germanium semiconductors.

The research done during the World War II period was carried out under strict secrecy regulations. Results were given orally and in "secret" reports only to those who officially had "a need to know". Late in 1945 the appropriate U.S. government agency declassified the basic research on semiconductors done by the Purdue group.

The first opportunity at which Lark-Horovitz could publicly present results from his war-time research was at the meeting of the American Physical Society held at Columbia University on January 24–26, 1946. An overflow crowd attended a Saturday afternoon session to listen to and ask questions about a related sequence of five 10-minute papers relating to basic electrical properties of germanium. The corresponding abstracts, four by Lark-Horovitz and colleagues and the fifth by Conwell and Weisskopf, form Selection 9 of Part 2.

During the fall of 1945 Lark-Horovitz decided that solid state physics offered to him a challenging and exciting field for his future research efforts. He had associated with him a substantial number of graduate students and senior colleagues to form a viable group to carry on research in this field. This group was shortly increased and strengthened by the addition of several senior physicists, notably Professors H. Y. Fan, A. N. Gerritsen, and P. H. Keesom. Lark-Horovitz welcomed financial support from the U.S. Army Signal Corps. He decided to concentrate particularly on basic research on the physical properties of semiconductors, especially elementary semiconductors. In 1946 he stated his goal as that of obtaining a clear picture of the behaviour of semiconductors, both from the structural and the electrical points of view, and to correlate experimental observations with theory.

Such an ambitious programme could not be developed by a single individual, and thus Lark-Horovitz continued his established practice of involving himself simultaneously in a half-dozen or more separate studies, each one carried out with a different associate or group of associates. During the succeeding years his interest remained focused on the behaviour of semiconductors but with increasing scope to the phenomena under observation.

An early theoretical success was the prediction that it is possible to trace the transition in the behaviour of the conduction electrons (or holes) in a semiconductor from the classical phase characterized by Maxwell–Boltzmann statistics to the quantum phase in which Fermi–Dirac statistics are required. This transition occurs when the density of accessible conduction band states decreases more rapidly, with fall in temperature, than does the conduction electron concentration. Previously one was restricted to observing either the classical behaviour alone, as in insulators or high-resistivity semiconductors, or the quantum behaviour only, as in metals. In semiconductors with a moderate value of the forbidden energy gap, such as germanium (0·72 eV), the transition is observed for low-resistivity samples as the temperature is reduced from room temperature to temperatures in the liquid hydrogen or liquid helium range. Lark-Horovitz was invited to present these results to the American Physical Society meeting held in Washington, D.C. in April 1947. The gist of this work is contained in the publication by Johnson and Lark-Horovitz which constitutes Selection 10 of Part 2.

Application of this theoretical work led to predictions relating to the low-temperature resistivity behaviour of semiconductors. Because, in 1947, Purdue lacked facilities for obtaining temperatures lower than that of liquid nitrogen, the knowledge of the experimental behaviour of germanium in the 4–20°K range was obtained from measurements made at Carnegie Institute of Technology (I. Estermann and students) and at the Naval Research Laboratory (J. Ambrose). By August 1949 a Collins helium liquefier was operational at Purdue and investigations of the physical properties of solids at low temperatures were started by P. H. Keesom, N. Pearlman, and their associates.

As early as 1942 Lark-Horovitz had initiated experimental investigations of the resistivity, Hall coefficient, and thermoelectric power of semiconducting samples. He also started, shortly thereafter, theoretical studies whereby experimental results could be converted into knowledge of the conduction electron behaviour. Following the end of the war, these theoretical studies were continued. Selection 11 of Part 2 serves as an example of this type of study; the theory of thermoelectric power in semiconductors was developed so that values could be calculated from resistivity and Hall coefficient data and then compared with measured thermoelectric data.

As a result of his experience with structure analysis by X-ray techniques, Lark-Horovitz applied these methods to various semiconductors. He firmly believed that a knowledge of the structure is necessary to the understanding of total physical behaviour of a material. In 1946 he and Dowell carried out an X-ray investigation of a graded sequence of polycrystalline germanium samples doped with differing amounts of tin. They could discriminate between the tin taken into the germanium lattice and the free tin deposited on the grain boundaries of the samples. The electrical conductivity behaviour correlated with the X-ray information concerning the amount of free tin present.

During 1954–6 Lark-Horovitz, Buschert, and Geib carried out X-ray structure analyses of molten semiconductors. The structure of solid tellurium is explained on the basis of a chain structure characterized by a bond distance of 2·86 Å. These chains were found to be still present in molten tellurium at 465°C (m.p. = 450°C) and to persist to some extent in tellurium at 610°C. Crystalline selenium, another elementary semiconductor, has the same structure as tellurium in the solid phase, and, just as with tellurium, the X-ray studies of selenium showed the continuation of the chain structure in the molten phase at 235°C and 310°C (m.p. = 217°C). A similar investigation of molten indium antimonide at 540°C (m.p. = 523°C) showed that, for this substance, the tetrahedral structure of the solid is not retained in the liquid.

Next to germanium, the elementary semiconductor to which

Lark-Horovitz gave the most attention was tellurium. In 1947 Lark-Horovitz and Scanlon studied the electrical behaviour of tellurium films evaporated in a high vacuum onto a mica substrate and obtained inconclusive results. They followed this with a study of the relative importance of galvanomagnetic and thermomagnetic effects in bulk polycrystalline tellurium. In 1950–1 Lark-Horovitz and Bottom carried out a detailed study of the electrical properties of tellurium, including determination of the forbidden energy gap, investigation of the two sign reversals of the Hall coefficient as temperature is changed appropriately, and an initial study of the anisotropy of resistivity and Hall coefficient in single-crystal material. Finally, in 1957 Lark-Horovitz, Epstein, and Fritzsche published the results of their research on the electrical properties of tellurium at the melting point and in the liquid state. They found that semiconducting properties persist into the liquid state and that a gradual transition to metallic conduction occurs as the temperature rises.

Goldsmith and Lark-Horovitz studied the photoconductive properties of cadmium sulfide and showed in 1949 that single crystals of CdS could be used as alpha, beta, and gamma counters. About the same time Lark-Horovitz, Fan, Orman, and Goldsmith found that germanium *p–n* barriers could be used as alpha and beta counters.

Owing to his background as a structure physicist, Lark-Horovitz considered electron microscope investigation of the surface of samples to be a fruitful technique. In 1950, with Daughty, Roth, and Shapiro, he carried out X-ray and electron diffraction studies of films of PbS, PbSe, and PbTe. These films were obtained by evaporation of the compounds onto glass, mica, or quartz; in addition, PbS films were deposited on the same surfaces from solution. An interesting result was that the patterns formed by PbS films heat-treated in air corresponded not only to PbS, but also to lanarkite ($PbO.PbSO_4$).

In 1951 Thornbill and Lark-Horovitz studied germanium films deposited in a high vacuum onto quartz and glass to thicknesses of 0·5 to 7 microns. X-ray diffraction indicated the normal diamond lattice. All films were found to be *p*-type with very low carrier mobilities, indicating a high degree of imperfection. Lark-Horovitz and W. M. Becker in 1952 found that thin germanium films deposited on quartz by the thermal dissociation of GeH_4 all showed *p*-type conductivity with carrier mobilities lower than usually observed in bulk germanium samples. In 1955 they prepared thin germanium layers a few microns thick by grinding and etching single crystals of bulk germanium. These layers were shown to still be single crystals by X-ray analysis, and they resembled in mobility and Hall coefficient sign the bulk samples from which they originated. Further experiments by Kellett, Fritzsche, and Lark-Horovitz confirmed the result that germanium layers formed by grinding and etching bulk single crystals down to a thickness of 12 microns exhibit, between 1·3 and 300°K, the same temperature behaviour of resistivity and Hall coefficient as does the bulk, provided that the layers are free of strain.

In 1949 investigations of the photoeffect and photoconductivity of the *p–n* boundary led to some startling discoveries in the optics of semiconductors. It was found at this time that the infrared transmissivities of pure germanium and pure silicon are very high indeed at wavelengths longer than the values corresponding to the absorption edges. The infrared transmission measurements were made by M. Becker and H. Y. Fan [*Phys. Rev.* **76,** 1530, 1531 (1949)]. Lark-Horovitz and K. W. Meissner collaborated by measuring the reflectivity of germanium samples, both *p*-type and *n*-type, high and low resistivity, by the method of residual rays in the range from 8·7 to 152 microns.

Lark-Horovitz extended his war-time work on metal–semiconductor rectifying contacts in work with R. Bray on photoconductive and photovoltaic effects produced on metal–*p*-type Ge rectifiers. They also investigated the effect of strong electric fields on the observed spreading resistance of metal–germanium point-contact rectifiers. Some of these experiments closely resembled the work carried out simultaneously at the Bell Telephone Laboratories by Bardeen, Brattain, Shockley and associates as they studied carrier injection and invented the transistor.

During 1949 and 1950 Lark-Horovitz, Taylor, and Roth succeeded in developing techniques for the routine production of large single crystals of high-purity, systematically doped germanium.

In 1950 Hung and Gliessman [*Phys. Rev.* **79,** 726 (1950)] had observed that the electrical behaviour of germanium shows anomalies at temperatures in or near the liquid helium range. Specifically, they found that, at first, resistivity and Hall coefficient increase exponentially with decreasing temperature as expected from the simple theory of elementary semiconductors, but, at a low temperature characteristic of the sample, the Hall coefficient reaches a maximum and then decreases sharply with further temperature drop. In the same temperature range the resistivity reaches a saturation value. Fritzsche and Lark-Horovitz carried out a more quantitative study of this effect in single-crystal *n*- and *p*-type germanium. They added transverse magneto-resistance to the group of electrical properties considered and proposed a model which assumes conduction in two bands, the regular conduction band (valence band for *p*-type material) and a band characterized by a very small mobility (impurity band). Conduction in the impurity band is responsible for the observed anomalous behaviour at very low temperatures. Selection 12 of Part 2 includes this paper by Fritzsche and Lark-Horovitz and a related later paper by Longo, Ray, and Lark-Horovitz on impurity band conduction in silicon.

After the introduction of III–V semiconducting compounds by Welker in 1952, Lark-Horovitz extended his own work to include these new and interesting substances. In 1955 Fritzsche and Lark-Horovitz reported on their studies of the electrical properties of *p*-type single crystals of indium antimonide in the temperature range from 270°K down to 1·5°K. The low-temperature behaviour of resistivity and Hall coefficient was anomalous in the same manner as observed in germanium and silicon. Thus InSb could be characterized as showing impurity band conduction at sufficiently low temperature. In addition, the transverse magneto-resistive ratio was found to change sign from + to − as the sample is cooled below the temperature at which the low-temperature maximum in the Hall coefficient occurs.

Out of all of his activities in physics, and especially in solid state physics, it is undoubtedly true that Karl Lark-Horovitz is best known for his many investigations into the effects of various kinds of radiation upon the properties of semiconductors. He and his colleagues initiated this field of research, with their first published reports appearing in 1948. Selection 13 of Part 2 is composed of abstracts of the initial work on the deuteron bombardment and neutron irradiation of germanium.

Selection 14 of Part 2 contains four contributions by Lark-Horovitz and colleagues, all published as letters to the Editor of the *Physical Review* in 1949 and 1950. These give a more complete picture of the early results obtained by the neutron irradiation of germanium and silicon.

During the summer of 1950 a conference on semiconducting materials was held at the University of Reading under the auspices of the International Union of Pure and Applied Physics in co-operation with the Royal Society. The only paper given there on irradiation effects was the presentation of Lark-Horovitz on nucleon-bombarded semiconductors. This landmark work constitutes Selection 15 of Part 2.

At first only heavy particles (p, d, n, α) were used in the irradiation studies. In 1951 Klontz and Lark-Horovitz announced their first results obtained from the bombardment of germanium with electrons. They discovered a threshold in the 0·5–0·7 MeV range; electrons of this energy or higher produced lattice displacements in germanium. These displacements were detected by changes in the rectification characteristics, conversion of n-type material to p-type, and increase in the conductivity of p-type germanium.

By 1951 rather extensive experimental information had been obtained about the effects produced through the bombardment of germanium and silicon with heavy particles or electrons. In an attempt to develop a more complete basis for the understanding of the properties of bombardment semiconductors, James and Lark-Horovitz developed a theoretical treatment of the localized electronic states produced in bombarded semiconductors. This may be found as Selection 16 of Part 2.

Selection 17 presents the results obtained by Lark-Horovitz and a group of associates working at the Oak Ridge National Laboratory in their study of the effects produced by the fast neutron bombardment of germanium. These experiments are interpreted in the light of the theory of James and Lehman which is given as an Appendix in Selection 15.

In the bulk of the research on irradiation effects the measures used were changes in electrical properties like electrical conductivity and Hall coefficient, i.e. properties which respond to changes in charge carrier concentration or the introduction of additional scattering centres. An exception to this research routine was provided by Keesom, Lark-Horovitz, and Pearlman when they measured the effect of neutron bombardment on the low-temperature atomic heat of silicon (Selection 18).

During 1952–4 Lark-Horovitz collaborated with Forester and Fan in studying the results of 9·3 MeV deuteron bombardment of *n*- and *p*-type germanium and *p*-type silicon. He also worked with Klontz, MacKay, and Pepper in extending the knowledge of electron irradiation effects in germanium. The results of this work were combined with results obtained from alpha particle irradiation of germanium, and the general problem of fast particle irradiation of germanium semiconductors was analysed from the theoretical viewpoint. This study was carried out by Fan and Lark-Horovitz and presented in 1954 to the international conference on Defects in Crystalline Solids held at Bristol, England. This paper constitutes the nineteenth and last selection of Part 2.

During the last three years or so of his lifetime Lark-Horovitz started a number of promising young men on research programmes related to the general programme of irradiation studies. Generally speaking, these studies were completed and published by the students after the death of Lark-Horovitz in 1958. A glimpse into his wide range of activity is provided by noting that he was simultaneously directing Ph.D. thesis research by L. Aukerman on electron bombardment of single-crystal specimens of III–V compound semiconductors, D. Kleitman on the deuteron irradiation of *n*-type germanium, G. W. Gobeli on the alpha-particle irradiation of germanium at 4·2°K, and D. E. Hill on electron bombardment of single-crystal silicon. And, simultaneously, he was working, as mentioned above, with Epstein and Fritzsche on molten germanium and tellurium and with Longo, Ray, and Fritzsche on impurity band conduction. He discussed enthusiastically other facets of semiconductor behaviour with senior colleagues and graduate students and offered many cogent suggestions relative to the puzzling problems that arose.

Epilogue

DURING the summer of 1958 H. M. James, the successor of Karl Lark-Horovitz as head of the Purdue University Department of Physics, wrote as follows:

"In April the Physics Department suffered a great loss through the death of Dr. Karl Lark-Horovitz, who was Director of the Physical Laboratory for thirty years, and had been Head of the Department since 1932. During this period the energy and abundance of ideas of Professor Lark-Horovitz had completely changed the character of the Department, and had broadened and strengthened its contribution to the work of the University. . . . His many graduate students would agree that he was an exacting taskmaster, but one who was extremely fertile in ideas, who had an exceptional familiarity with the pertinent literature, and who succeeded in communicating some of his own vigour and enthusiasm to them. . . . In memory of his role in the growth and development of the Department of Physics, the Lark-Horovitz Prize in Physics† has been endowed by the gifts of his family, the staff of the Department, former students, and other friends and associates."

Finally, President-Emeritus E. C. Elliott, the man who brought Lark-Horovitz to Purdue and then backed him fully in his efforts at building a department of recognized high standing, wrote these words on April 21, 1958 to be read at an informal memorial meeting held by the members of the Department:

"Two or three days prior to the tragic end, it was a great personal satisfaction for me to have a somewhat lengthy visit with

† Awarded annually to the Ph.D. candidate in the Department whose research is judged to be the best by a committee of professors representing the various fields of physics.

Dr. Lark-Horovitz. Remembering that I had been away from the University most of the time in the last dozen years, I was amazed at the progress of the Department which he outlined for me.

"Karl Lark-Horovitz was an educated scientist concerned not only with the subject of physics, but all collateral knowledge. He was, furthermore, concerned with the teaching of his subject to young students. I was most fortunate that he served the University at a critical stage when the unity of the sciences was a matter of concern for those charged with the responsibility for leadership. During the years that I worked with him, I found him most cooperative, though at times most impatient on account of the slowness with which things were moving.

"The University has a Department of Physics of which any institution might be proud. This we owe largely to him who has recently gone from us. It gives me great satisfaction to have a small part in this memorial service for one who endowed the University with his knowledge, energy and personality for thirty years."

List of Publications

1. The historical development of physical concepts of relativity (in German), *Arch. Gesch. Naturw. Tech.*, **5,** 251 (1914).
2. On absorption measurements with radioactive elements (in German), *Sitz.-ber. Akad. Wiss. Wien*, Abt. IIa, **123,** 1819 (1914), with F. Paneth.
3. On absorption measurements with radioactive elements (in German), *Zeits. f. phys. Chemie* **89,** 513 (1915), with F. Paneth.
4. On the apparent diminution of objects travelling away from the observer (in German), *Phys. Zeits.* **21,** 499 (1920).
5. Contribution to the theory of the field of vision (in German), *Sitz.-ber. Akad. Wiss. Wien*, Abt. IIa, **130,** 405 (1921).
6. Investigations on the theory of the field of vision (in German), *Ber. der deutsch. phys. Ges.*, Verhandlungen, Reihe 3, Jahr 2, S. 9 (1921).
7. Size perception and depth of vision (in German), *Pflug. Arch. für Physiol.* **194,** 629 (1922).
8. Heteromorphies due to the variation of effective aperture and visual acuity, *J. Opt. Soc. Amer.* **6,** 597 (1922).
9. The hydrogen-electrode function of platinum (in German), *Sitz.-ber. Akad. Wiss. Wien*, Abt. IIa, **132,** 367 (1923).
10. The determination of crystal structure by using radioactive materials (in German), *Sitz.-ber. Akad. Wiss. Wien*, Abt. IIa, **132,** 375 (1923).
11. Ion exchange in dielectrics. I. Electrode functions of glasses (in German), *Zeits. f. Physik* **15,** 369 (1923).
12. Characterization of glasses by their electromotive properties

(in German), *Sitz.-ber. Akad. Wiss. Wien*, Abt. IIa, **134,** 335 (1925), with F. Horn, J. Zimmermann, and J. Schneider.
13. On the formation of mixed electrodes at phase boundaries (in German), *Zeits. phys. Chemie* **115,** 424 (1925).
14. Investigations of ion exchange in glasses (in German), *Sitz.-ber. Akad. Wiss. Wien*, Abt. IIa, **134,** 355 (1925), with Josef Zimmermann.
15. A focussing X-ray spectrograph for low temperatures, *Science* **64,** 303 (1926).
16. A permeability test with radioactive indicators, *Nature* **123,** 277 (1929).
17. The phase-boundary potential difference at the boundary between a dielectric and its water solution (in German), *Naturwiss.* **19,** 397 (1931).
18. Electromotive force of dielectrics, *Nature* **127,** 440 (1931).
19. An X-ray investigation of oriented dielectrics, *Proc. Indiana Acad. Sci.* **43,** 182 (1934), with A. N. Ogden.
20. Structure of the wood used in violins, *Nature* **134,** 23 (1934), with W. I. Caldwell.
21. The structure of wood in violins (in German), *Naturwiss.* **22,** 450 (1934), with W. I. Caldwell.
22. A new method of making extremely thin films, *Rev. Sci. Instr.* **6,** 401 (1935), with J. D. Howe and E. M. Purcell.
23. The origin of the "extra rings" in electron diffraction patterns, *Phys. Rev.* **48,** 101 (1935), with H. J. Yearian and J. D. Howe.
24. Intensity distribution in electron diffraction patterns, *Proc. Amer. Phil. Soc.* **76,** 766 (1935), with H. J. Yearian and J. D. Howe.
25. A simple method for testing homogeneity of wood, *Nature* **137,** 663 (1936).
26. X-ray analysis of polymorphous mixtures, ASTM Symposium on Radiography and X-ray Diffraction, p. 298 (1936).
27. Physics teaching and text books, *Science* **88,** 354 (1938).
28. Nuclear excitation of indium with alpha particles, *Phys. Rev.* **55,** 878 (1939), with J. R. Risser and R. N. Smith.

29. Intensity distribution in X-ray and electron diffraction patterns, *Phys. Rev.* **57,** 603 (1940), with C. H. Ehrhardt.
30. Structure of liquid argon, *Nature* **146,** 459 (1940), with E. P. Miller.
31. Uranium fission with Li–D neutrons: energy distribution of the fission fragments, *Phys. Rev.* **60,** 156 (1941), with R. E. Schreiber.
32. Electron diffraction intensities, *Nature* **148,** 287 (1941), with H. J. Yearian.
33. The intake of radioactive sodium and potassium chloride and the testing of enteric coatings, *J. Applied Phys.* **12,** 317 (1941).
34. Radioactive indicators, enteric coatings, and intestinal absorption, *Nature* **147,** 580 (1941), with H. R. Leng.
35. A new method of testing enteric coatings, *J. Amer. Pharmaceutical Assoc.* **31,** 99 (1942), with H. R. Leng.
36. Report of the committee on the teaching of physics in secondary schools, *Amer. J. Physics* **10,** 60 (1942).
37. Shop work for the physics teacher, *Amer. J. Physics* **10,** 161 (1942).
38. High school science and mathematics in relation to the manpower problem, *School Science and Mathematics* **43,** 127 (1943), with R. L. Havighurst.
39. Science in the schools of tomorrow, *School Science and Mathematics* **43,** 64 (1943).
40. On the preparation and certification of teachers in secondary school science, *Amer. J. Physics* **11,** 41 (1943).
41. The schools in a physicist's war, *Amer. J. Physics* **11,** 102 (1943), with R. J. Havighurst.
42. Screening for training, *J. Amer. Assoc. University Women,* p. 1 (1943).
43. On the teaching of the basic sciences, *Amer. J. Physics* **12,** 359 (1944), with K. W. Bigelow, R. J. Havighurst, and F. J. Kelly.
44. The preparation of high school science and mathematics teachers, *School Science and Mathematics* **46,** 107 (1946).

45. Responsibilities of science departments in the preparation of teachers, *Amer. J. Physics* **14,** 114 (1946).
46. Transition from classical to quantum statistics in germanium semiconductors at low temperature, *Phys. Rev.* **71,** 374, 909 (erratum) (1947), with V. A. Johnson.
47. Alpha-2 neutrons nuclear reactions, *Phys. Rev.* **72,** 1117 (1947), with J. R. Risser and R. N. Smith.
48. Effectiveness of science teaching, *Science Teacher*, Feb. 1948.
49. A chronology of scientific development, 1848–1948, compiled for the centennial celebration of the AAAS, published by the AAAS (Sept. 1948), with Eleanor Carmichael.
50. Cadmium sulfide as a crystal counter, *Phys. Rev.* **75,** 526 (1949), with G. J. Goldsmith.
51. Neutron irradiated semiconductors, *Phys. Rev.* **76,** 442 (1949), with W. E. Johnson.
52. Conductivity in semiconductors, *Electrical Engineering* **68,** 1047 (1949).
53. The optical properties of semiconductors. I. The reflectivity of germanium semiconductors, *Phys. Rev.* **76,** 1530 (1949), with K. W. Meissner.
54. Science for the non-scientist, *Main Currents of Modern Thought* (Fall 1949).
55. Transmutation-produced germanium semiconductors, *Phys. Rev.* **78,** 814 (1950), with J. W. Cleland and J. C. Pigg.
56. Report of the cooperative committee for the teaching of science: a report to the AAAS Council, *Science* **111,** 197 (1950).
57. Fast neutron bombardment effects in germanium, *Phys. Rev.* **78,** 815 (1950), with J. H. Crawford, Jr.
58. Electron mobility in germanium, *Phys. Rev.* **79,** 409 (1950), with V. A. Johnson.
59. Thermal equilibrium in neutron-irradiated semiconductors, *Phys. Rev.* **79,** 889 (1950), with J. H. Crawford, Jr.
60. Theoretical Hall coefficient expressions for impurity semiconductors, *Phys. Rev.* **79,** 176 (1950), with V. A. Johnson.

61. Nucleon-bombarded semiconductors, *Semiconducting Materials*, p. 47, Butterworths Scientific Publications (1951).
62. Localized electronic states in bombarded semiconductors, *Zeits. f. phys. Chemie* **198,** 107 (1951), with H. M. James.
63. The combination of resistivities in semiconductors, *Phys. Rev.* **82,** 977 (1951), with V. A. Johnson.
64. The effect of fast neutron bombardment on the electrical properties of germanium, *Phys. Rev.* **83,** 312 (1951), with J. W. Cleland, J. H. Crawford, Jr., J. C. Pigg, and F. W. Young, Jr.
65. Evidence for production of hole traps in germanium by fast neutron bombardment, *Phys. Rev.* **84,** 861 (1951), with J. W. Cleland, J. H. Crawford, Jr., J. C. Pigg, and F. W. Young, Jr.
66. The effect of neutron bombardment on the low temperature atomic heat of silicon, *Science* **116,** 630 (1952), with P. H. Keesom and N. Pearlman.
67. Semiconducting films, *Proc. of National Electronics Conf.* **8,** 506 (1952), with W. M. Becker.
68. Theory of thermoelectric power in semiconductors with application to germanium, *Phys. Rev.* **92,** 226 (1953), with V. A. Johnson.
69. Energy levels and photoconductivity in electron-bombarded germanium, *Phys. Rev.* **95,** 1087 (1954), with H. Y. Fan, W. Kaiser, E. E. Klontz, and R. R. Pepper.
70. The electrical properties of germanium semiconductors at low temperatures, *Physica* **20,** 834 (1954), with H. Fritzsche.
71. The new electronics, *Present State of Physics*, p. 57, AAAS, (1954).
72. Fast particle irradiation of germanium semiconductors *Report of the Conference on Defects in Crystalline Solids, Bristol*, p. 232, published by Physical Society (London), 1955, with H. Y. Fan.
73. Electrical properties of *p*-type indium antimonide at low temperatures, *Phys. Rev.* **99,** 400 (1955), with H. Fritzsche.
74. Electrical properties of tellurium at the melting point and in

the liquid state, *Phys. Rev.* **107,** 412 (1957), with A. S. Epstein and H. Fritzsche.

75. The physics of semiconductors, *The Science of Engineering Materials*, p. 336, ed. by J. E. Goldman, Wiley, 1957, with V. A. Johnson.
76. Semiconductors at low temperatures, *Progress in Low Temperature Physics*, Vol. 2, p. 187, ed. by C. J. Gorter, North-Holland Publishing Co., Amsterdam, 1957, with V. A. Johnson.
77. On the photoconduction of germanium after irradiation with fast electrons (in German), *Zeits. f. Physik* **153,** 331 (1958), with F. Stockmann, E. E. Klontz, J. MacKay, and H. Y. Fan.
78. Low temperature impurity conduction in silicon, *J. Phys. Chem. Solids* **8,** 259, (1959), with T. A. Longo and R. K. Ray.
79. Effect of minority impurities on impurity conduction in *p*-type germanium, *Phys. Rev.* **113,** 999 (1959), with H. Fritzsche.
80. Irradiation of semiconductors, *Semiconductors and Phosphors*, Proceedings of the International Colloquium held at Garmisch-Partenkirchen in 1956, Vieweg, Brunswick, 1958, with H. Y. Fan.
81. Irradiation effects in semiconductors, *Effects of Radiation on Materials*, p. 159, Reinhold, 1958, with H. Y. Fan.
82. Solid state physics, *Methods of Experimental Physics*, Vol. 6A, p. 14, Academic Press, N.Y., 1959, with V. A. Johnson.
83. K. Lark-Horovitz and V. A. Johnson, Editors, *Methods of Experimental Physics*, Vols. 6A and 6B, Solid State Physics, Academic Press, N.Y., 1959.

The *Physical Review* (1926–55) and the *Bulletin of the American Physical Society* (1956–8) contain 91 abstracts for which Lark-Horovitz is listed as the author or co-author.

PART 2

HIS WORK

1

Heteromorphies Due to the Variation of Effective Aperture and Visual Acuity†

1. By the use of optical apparatus the image of the surrounding space sometimes undergoes a complete change, as in some kinds of prisms and mirrors. But also in other cases, in which the instrument produces a perfect image, apparent alterations of space are perceptible. These apparent defects of the visual space are called heteromorphies.

Expert microscopists are always surprised that beginners are unable at first to find the image, or, when drawing it, always make it too small. And to each of us it is well known that the drawing or photograph of a microscopical object seems much bigger than really shown by the microscope. Impressions of this kind are also perceptible in other cases than when using an optical instrument. If, for instance, instead of through the microscope we look through an empty tube with one eye and fix it on a distant object[1] while the other eye is directed to the same object in the ordinary manner, the image seen through the tube appears much smaller than to the naked eye.[2] Exactly the same thing can be observed by looking at a landscape through the finder of a camera or by holding a narrow stop before the eye. In all these cases the aperture of the rays of the optical system, consisting of the eye together with the effective stop, is changed.

The phenomena described can be observed by persons with normal vision (emmetropes), under certain conditions to be described later, by ametropes and, as already mentioned, in using

† *J. Opt. Soc. of Am. and Rev. Sci. Instr.* **6**, 597 (1922), communicated by Dr. L. Silberstein.

optical instruments. The phenomena are therefore, as it seems, independent of the dioptrical properties of the system, but only so far as the dioptrical system remains unchanged and the image remains clear.

2. The effective aperture may influence the image in different ways. When the entrance-pupil is altered, the size of the image-forming diffraction-disks is also altered. The smaller the latter are, the sharper the images, but their brightness is reduced (case of a small diaphragm). A stop which reduces the aperture of the image-forming pencils without altering the entrance-pupil changes the field of view. The image of the stop in the focal plane (in the case of the eye, on or before the retina) is the exit-window (case of the tube and the finder of a photographic camera).

An alteration of the entrance-pupil changes also the actual size of the image, for the circles of confusion become smaller. As the aberrations in the eye and the formation of the image by wide-angle pencils bring it about the size of the diffraction-disks depends not only on the size of the entrance-pupil, this reduction of the image on the retina is difficult to perceive. Thus the diminution observed, when looking through a diaphragm, need not depend necessarily on the actual reduction of the image on the retina. When, the entrance-pupil remaining constant, the entrance-window or the stop of the field of view is changed, the structure of the image is only changed in those places where now, instead of the former images, the image of the exit-window is formed. If nevertheless the impressions brought about by the unchanged retinal images are changed, evidently this must be connected with the mutual influence of neighbouring parts of the retina,[3] especially with the influence of the simultaneous contrast-sensibility which brings about a change of the adjustment of the eye. Experiments have shown that the diminution is as much increased as the stop of the field of view decreases (that is to say, in the case of the tube it would be a longer one). A diminution of the field of view also occurs with the replacement of the binocular by monocular vision, which transition also is connected with an apparent diminution of size.

But an alteration of the aperture is also connected with an alteration of the distribution of light in the image. This is caused by a change of the circle of confusion by variation of the entrance-pupil. We have, further, to consider a change of the simultaneous contrast, when, by an alteration of the stop of the field of view, parts of the image are covered either by light or darkness. The former case occurs when an object is observed through a glass-tube, which, by total reflection, forms on the retina a brightly illuminated background; the latter when looking through a tube blackened on the inside, in which case the part of the field of view around the retinal image is darkened. But this affects in a decisive way the acuteness of vision and this influence therefore must always be taken into account.

3. It is well known that the acuteness of vision is diminished by a reduction of the intensity of illumination, by flooding the retina with useless light, by the darkening of simultaneous contrast and of course also by wrong adjustment.[4] All these produce a reduction of the relative differences in the sensations of the brightness of two points. An increase of the visual acuity is caused by: a diminution of the entrance-pupil, a moderate increase of the intensity of light (provided that the peripheral portions of the retina are not sensibly illuminated by stray light), and the illuminating simultaneous contrast. Each change of the incident light alters the pupillary diameter leading to disturbing secondary phenomena (the acuteness of vision may be changed thereby and also focusing movements be liberated).[5] Therefore, it was necessary to make experiments in which the influence of the size of the pupil was eliminated. For this purpose investigations were made on persons, whose pupils had lost the reaction to light. *The result was, that a reduction of the acuteness of vision is always followed by an apparent diminution in size and, at the same time, the removing and bringing nearer together of the objects observed.* On the other hand a sudden increase of the visual acuteness produces an increase of size. With atropinisized ametropes it could be observed that a small (stenopäical) stop which in the case of normal sight would cause a diminution improves the acuteness of vision so much that an

increase may be observed. Beside this *new* effect of the alteration of size in the field of vision, the known influence of the visual acuity must also be considered and also the influence of the aperture on the depth of focus of the image-space. The importance of the acuteness of vision for the resolving power of the microscope was pointed out by F. E. Wright.[6] (Here the illuminating effect of the simultaneous contrast comes into play.) In the following I intend to deal with a series of contrivances, for which the above mentioned effects are of importance, and then set forth the theory of the phenomena.

4. For the series of readings a simple lens is used, before which a diaphragm is placed, to increase the distinctness. With a very small diaphragm (e.g., of 0·1 mm diameter) the usual magnification cannot be observed. I made this observation myself with an Elster-Geitel-electroscope and found it confirmed by other observers: the scale and the leaves of the electroscope seem to be in one plane and far more distant than without a diaphragm. As these phenomena are not always perceptible with both eyes with the same intensity they are also of importance when first one eye is used and then the other in observing and above all in comparing the images with one eye aided and the other naked. This is the case with the well-known examination of the magnifying power of a telescope which consists in comparing the lines of a distantly suspended scale with one eye naked and the other aided by the telescope. If then the exit-pupil of the telescope is much smaller than the pupil of the eye, the phenomenon mentioned above is observed. Thus sometimes it is not sufficient to pull out the tube to diminish the parallax between the two images. A similar examination of the magnifying power of a microscope is open to the same difficulties. To avoid such faults, fix close to the instrument, whose magnifying power is to be examined, a diaphragm for the naked eye of the size of the exit-pupil of the instrument. The right perception of the increase, when using night-glasses, depends also on the correct proportion between the exit-pupil of the instrument and the eye-pupil. A. Gehlhoff[7] pointed out that the size of the field of view is also of psychological importance for

the resolving powers of these instruments. By employing drawing apparatus, as for instance the well-known camera lucida, the influence of the opening of the stop is remarkable. On using this instrument the scenery appears smaller on the drawing paper than to the naked eye. *If the stop is made gradually smaller* (as far as it is possible without making the illumination of the image so faint that it is no longer perceptible) *the image becomes smaller as well.* Qualitative tests, made with regard to this, have shown that the variation in size by stopping down the diaphragm from 3 mm to 0·1 mm is about 15%.

It is natural to take into account the variation of visual acuity by dazzling. Therefore every physicist or astronomer uses in exact measurements a dark eye-shield to cover the non-observing eye. Also blinding of this eye, as is easily proved, diminishes the acuteness of vision and for this reason apparently the magnification and the perception of depth for the observing eye. Therefore it seems necessary, not to change the conditions for the formation of image in the eye, in applying optical apparatus for subjective use.

5. According to the doctrines of physiological optics, the defective aperture of the rays and the acuteness of vision do not immediately determine the perception of size. But these factors do determine the depth of focus which depends on p/σ, where p denotes the diameter of the pupil and σ is the angular measurement of the visual acuity.[8] If p/σ diminishes, the depth of focus increases, and *vice versa.* The distinctness of the perception of depth is inversely as the optically defined range of distinct vision (depth of focus). On the other hand the sensation of size is connected with the accommodation of the eye. Any object seems smaller in proportion as the accommodation is greater or even when the sense of greater accommodation is excited, and this is the case although the image on the retina is unchanged in size. We will assume now, that if anywhere a variation of the range of distinct vision takes place, the accommodation or the innervation of accommodation increases as far as possible without perceiving the image less sharply (*Principle of maximum accommodation*).[9] If the range of distinct vision increases, a point is approached at

which the depth begins to be practically infinite and it is, therefore, useful to focus at a nearer point. This does not mean that these focusing impulses are always connected with a real accommodation, for we would then see the nearer point as sharply as the point at which we focused previously. On the contrary, impulses to relax the accommodation take place if the depth of focus decreases again because points at a greater distance are now also distinctly visible and thus the depth of the visual space is enlarged. By this conception, all the experiments mentioned above are intelligible: whether the pupil p or the angular size of visual acuity σ are changed,[10] the innervation to focus begins and together with it a variation of the impression of size and depth.[11]

6. To show how these changes of our optical sensations are connected with normal sight, we proceed to consider the conception that the space as seen by us is an optical transformation (in the sense of Maxwell and Abbe) of the physical space.

In this transformation any point of the object-space has a one to one correspondence with a point of the visual space and the lines of sight remain invariant, while the points at infinity are transformed into a plane at finite distance perpendicular to the axis of vision. Mathematically formulated these conditions give, instead of the usual optical transformation

$$x' = \frac{a_1x + a_2y + a_3z + a_4}{ax + by + cz + d}, \quad y' = \frac{b_1x + b_2y + b_3z + b_4}{ax + by + cz + d},$$

$$z' = \frac{c_1x + c_2y + c_3z + c_4}{ax + by + cz + d}$$

the equations:

$$x_v = \frac{a_1x}{ax + d}, \quad y_v = \frac{b_2y}{ax + d}, \quad z_v = \frac{c_3z}{ax + d},$$

with the condition[12] $b_2 = c_3 = d$. These are well-known formulae of projective geometry and give analytically the geometry of a relief-perspective for the general case. The origin of the coördinates lies in the first eye-point.[13] If $d = a_1$, the relief is the image of the real space which the *quiescent* eye sees, if the eye is contemplating the point at which it is focused (the point of view is identical with the first eye-point).[14] If the eye is focused on a point, which is nearer than that which is contemplated, then it is necessary for the restoration of the previous conditions to displace forward the eye until the point of view will be so near to the contemplated point, that the latter again is a focused one. But if the eye remains in the place of the first eye-point we must say that the point of view is shifted forward only *virtually:* the psychical adjustment of the eye and the innervation corresponds to a smaller distance. Then we have $d < a_1$, $(a_1 - d)/a$ being the range of the virtual displacement of the point of view, and the objects seem to be smaller and at a greater distance. Thus the conditions of variation of the visual space can be conceived by the rules given for normal vision.

For an exact theory it would be necessary to take into account these transformations only as a first approximation and to find the infinitesimal transformation which takes into account the apparent changes induced by the moving eye and the curvature on the margin of the field of vision.[15]

Summary

The influence of the effective aperture and the acuteness of vision have been dealt with in reference to well-known facts regarding the eye in connection with an optical instrument. It has been shown, that a reduction of the visual acuity entails a diminution of the apparent size. The importance of these facts for some optical observations and measurements is explained. This explanation is based on the assumption that maximum impulses to accommodate are liberated in such a manner that the distinctness of the image is not blurred when the conditions of the formation of the image are altered. The optical transformation of the object-space into the visual space for a quiescent eye is defined by simple postulates of invariance and is proved to be a relief-perspective. Usually the point of view coincides with the first eye-point. The heteromorphies considered above may be constructed as a virtual shifting forward of the point of view.

Notes

[1] It must be a distant object in order that the rays appear to come from infinity in the same manner as when using the microscope.

[2] Cf. St. Meyer, *Phys. Z.* **21,** 124 (1920); and K. Horovitz, *Phys. Z.* **21,** 499 (1920).

[3] This is Hering's induction.

[4] In the latter case also dioptrics may be of importance as mentioned before.

[5] Cf. E. Hummelsheim, *Arch. f. Ophth.* **45,** 357 (1898) and K. Horovitz, *Ber. d. deutschen physik. Ges.* **2,** 9–11 (1921) and *Sitz. Ak. Wiss. Wien,* **130,** 405–21 (1921).

[6] F. E. Wright, *J. Opt. Soc.* **2–3,** 101 (1919).

[7] A. Gehlhoff, *Z. f. techn. Phys.* **66,** 477 et seq.

[8] Cf. Rohr, *Brit. J. Phot.* **48,** 454 (1901) and S. Czapski, *Grundzüge der Theorie der optischen Instrumente*, pp. 256, 267 (1904).

[9] J. K. Horovitz, *Sitz. Akad. Wiss. Wien,* **130** (1921); Beiträge zur Theorie des Sehraums.

[10] It is necessary to mention, that a change of the pupil also entails a variation of visual acuity (but not *vice versa*).

[11] It is partly due to the influence of limiting the field of view, that the diminution of the object, when seen through concave spectacles, is much more intensely felt (as shown by Isakovitz), than the diminution of the image on the retina alone could bring about. Therefore, the variation of the depth of focus also must be considered: the entire depth being greater than without spectacles, because p/σ which determines the distinctness is only pB^2/σ ($B<1$).—It may be mentioned here, that also other dysmegalopsies, micropsy after the injection of atropin, macropsy after the use of pilocarpin (eserin) are intelligible if we assume the principle of maximum accommodation. In the first case the innervation is unlimited, the apparatus of accommodation being paralysed: hence micropsy. In the other case the external eye is in a spasm of accommodation, the impulses are arrested: hence macropsy.—For these and also other cases of physiological interest see K. Horovitz, *loc. cit.* and further an article appearing in *Pflüger's Archiv* (Grössenwahrnehmung und Sehraum-relief).

[12] As postulated by the invariance of the line of sight. It is not necessary for the visual space to be always symmetrical around the x axis, as in the case of astigmatism.

[13] Cf. Burmester, *Grundzüge der Reliefperspektive.*

[14] The case $d = a$ was developed by H. Witte, proceeding from experimental facts, without connection with the relief-formulae explained above. *Physik. Z.* **19** and **20,** several articles "Über den Sehraum".

[15] The present writer is pursuing these investigations in connection with Riemannian geometry. The observations stated here are qualitative and it would be most interesting to obtain some exact investigations on these points, which it was impossible to undertake here.

2

A Permeability Test with Radioactive Indicators†

CERTAIN investigators (see, for example, W. J. V. Osterhout, *Some Fundamental Problems in Cellular Physiology*, 1927, especially pp. 36–48) believe that the protoplasm of the living cell is permeable only to undissociated molecules but impermeable to ions.

It seemed possible to me to test this theory with the method of radioactive indicators[1] (Hevesy–Paneth). The advantage of this method is that only very small amounts of the ions which enter the cell are necessary and that a very small concentration can be detected. Radioactive lead (thorium-B) was used as an indicator for lead ions, and therefore lead nitrate was dissolved in sea water so as to make it 10^{-5} to 10^{-6} M in respect to lead ions. Cells of *Valonia macrophysa* were used since the large volume and the amount of sap available make the investigation easier, and since investigations of the permeability of this cell were carried out by Osterhout and his collaborators.

To test whether or not the presence of lead causes any injury to the cell, the cells were placed in sea water with different amounts of lead nitrate added, and for several months the behaviour of the cells observed. The cells did not change in colour or rigidity, and were, according to Dr. L. R. Blinks, who kept them in the same laboratory with other cells, in a normal state, judged from macroscopic appearance.

For the permeability experiments, the cells were placed in sea water containing a known amount of lead nitrate and thorium-B.

† *Nature* **123**, 277 (1929).

After 20 or 30 hours the cells were taken out, washed off with inactive sea water, and dried on blotting paper. The sap was removed, a certain amount (0·2–0·3 cc) evaporated in a watch glass, and the radioactivity measured in an α-ray electroscope. The activity of the same amount of the original solution and of the sea water in which the cells were kept was measured. In this way we ascertained how much lead is absorbed by the cell wall and how much enters the vacuole. In all experiments (14 cells) it was found that about 50% of the lead ions present in the original solutions are absorbed by the cell wall, but that practically no lead enters the vacuole.[2] The same experiments were carried out with cells which had been kept in sea water plus lead nitrate for four months. Also in this case no lead could be found in the vacuole.

One may conclude that all the lead which disappears from the sea water is adsorbed by the cell wall or the protoplasm forming an insoluble compound which cannot enter the vacuole. In this case one would expect that in dead cells also the lead would be fixed at the cell walls and therefore cannot be found in the sap. Experiments with three dead cells have shown that lead does enter a dead cell. It is apparently fixed there to small particles of organic matter which are to be found always in dead cells. Therefore it cannot diffuse back into the surrounding sea water and *an apparent concentration* of lead in the dead cells takes place.

It was interesting to see whether radium emanation, being a rare gas, would enter the cells, as one would expect from the theory. Small capillaries (16 mm long), filled with radium emanation (about 0·01 mc), were broken under the sea water containing the cells to be tested. It was found that already after one hour the sap is approximately as active as the surrounding sea water (15 cells were investigated).

After every experiment, Dr. L. R. Blinks examined the macroscopic appearance of the cells and tested the sap for sulphate ions. (The presence of sulphate ions would indicate a severe injury.) Part of the sap in our lead experiments and the sap of every single cell in the experiments with radium emanation was tested in this way. Injury was found in one cell out of a total of three, exposed for 20 hours in radium emanation, and traces of sulphate ions in two cases out of twelve, after 1 to 2 hours exposure in radium emanation. One cell that had been in lead nitrate for four months was soft, but did not give any sulphate reaction and did not show any sign of injury in our test.

SUMMARY

Using radioactive indicators for testing the permeability of single cells of *Valonia macrophysa*, it was found that lead ions do not enter the sap of the living cell even if the cells are kept for several months in lead nitrate solution. Lead ions enter readily the sap of dead cells. Radium emanation, being a rare gas, is already after one hour distributed evenly between the cell sap of living cells and the surrounding sea water.

This investigation was carried out in the spring of 1927 during our stay at the Rockefeller Institute for Medical Research, New York City, and we are indebted to the International Education Board who made our stay at the Rockefeller Institute possible.

NOTES

[1] That is, to determine the amount of ions present, of a certain kind, by the determination of the radioactive isotope mixed with them. Since a chemical separation of isotopes is impossible, the change in activity of the radioactive isotopes is the indicator for changes in the concentration of the inactive ion.

[2] A trace of activity which was found twice immediately after drying is due to traces of thorium-C. This may have entered the cell in ionic form, but since thorium-C is present only in an extremely small concentration, this is not contradictory to any other experiment on permeability. Such a small amount may possibly also enter in other cases, but could not be detected. On the other hand, thorium-C shows in neutral solutions a quasi-colloidal behaviour and may have entered the cell in form of an undissociated complex.

3

Electromotive Force of Dielectrics†

I FOUND some time ago[1] that glasses in aqueous solutions show the electromotive force of a solid electrolyte: the ion in the solid glass determines the potential difference against the solution. Moreover, it could be shown that cations from the solution were exchanged against the cations in the glass: the glass behaves then like a mixed electrode, but in a certain range of concentration practically like an electrode reversible to the ions taken up from the solution. Glasses which show definitely the behaviour of a solid electrolyte (sodium electrode) also show always exchange electrodes, particularly the silver and hydrogen electrode. Certain soft glasses show only the hydrogen electrode except in alkaline solution.

It was of interest to investigate whether other dielectrics would also show the same behaviour. With J. Hafner and lately with J. E. Ferguson, I investigated the electromotive behaviour of fused silica. In this case also we found the existence of the sodium, silver, and hydrogen electrode.

In the case of quartz, as in the case of glasses, the electromotive behaviour corresponds to the observations made when a current passes the solid.

J. E. Ferguson and I[2] have investigated during the last year the electromotive behaviour of thin paraffin films. We found also in this case the existence of sodium, potassium, hydrogen, and silver electrodes, and also (with some slight deviations, however) the existence of a calcium electrode, which we failed to find in any of the glasses investigated.

† *Nature* **127**, 440 (1931).

These phenomena can be understood[3] if we make the following assumption: The number of places available for cations in the dielectric is limited and constant ($= a$). Only one kind of ion, cations, are taken up by the solid, and only these ions can migrate in the solid. In the state of equilibrium the difference of potential, solid–solution, is the same for all kinds of ions present. Neglecting the number of ions given off by the solid as compared with the concentration, c, in the solution, and treating the solid phase in first approximation like a dilute solution, the following formula is obtained for two kinds of ions present:

$$E = \frac{RT}{F} \ln \frac{\frac{K_1}{u_1} u_2 a}{c_1 + \frac{K_1 u_2}{K_2 u_1} c_2}$$

and ΔE the potential difference in the "concentration cell" which alone is being measured:

$$\triangle E = \frac{RT}{F} \ln \frac{c_1' + A c_2}{c_1 + A c_2} \left(A = \frac{K_1 u_2}{K_2 u_1} \right).$$

In this formula c_1 are the concentrations in the solution, u_1 are the mobilities in the solid phase, K_1 are the integration constants in the expression for the thermodynamic potential of the ions present in the solid phase (solution tension).

In this formula all the terms can be measured directly, since also A is given by a single experiment. For permutites and certain glasses, more complicated formulae have to be used, since the ions in the solid phase cannot be treated as independent of one another, and since the amount of ions exchanged is comparable with the concentration of the ions in the solution.

NOTES

[1] *Z. Physik.* **15,** 1923.

[2] *Bull. Am. Phys. Soc.*, Nov. 1930.

[3] This formula has been used by me since 1924 for the interpretation of different experimental results. See, for example, *Bull. Am. Chem. Soc.*, April 1927, and J. W. V. Osterhout, *Bull. Nat. Res. Council*, **69,** 193, footnote 52, 1929.

4

Structure of Liquid Argon†

MOST of the X-ray investigations of liquids have led to a distribution function which cannot be compared directly to the corresponding distribution of atoms in the solid since one peak in the liquid distribution function usually corresponds to several peaks in the distribution function of the solid. For this reason we investigated several years ago the structure of molten salts such as potassium chloride and lithium chloride where one can assign to each peak in the liquid distribution function one peak in the distribution function of the solid. We have shown that the coordination in the liquid and the solid is the same; number of first neighbours N_1 in the solid equals 6 and in the liquid 5·8. The number of second neighbours N_2 in the liquid (K–K or Cl–Cl) is already greatly disturbed; in the solid there are 12 and in the liquid 9·8.

Since the development in recent years of a more refined theory of liquids[1] has permitted the prediction of the number and position of nearest neighbours in a normal liquid, it was of particular interest to study the structure of liquid argon.

Because liquid argon at atmospheric pressure is only stable over a temperature range of 3·5° (boiling point −185·7°C and melting point −189·2°C) a special container for the liquid was constructed which made it possible to apply pressures slightly above atmospheric pressure during the exposure.

Pure gaseous argon is introduced into the cell and is there condensed under a pressure of 910 mm mercury and at a temperature of 89·2°K. The liquid argon is held in a layer of about 1·3 mm thickness. The mica windows are sealed with indium gaskets

† *Nature* **146,** 459 (1940), with E. P. Miller.

to the copper walls of the cell proper which is cooled by a continuous stream of liquid air. This arrangement permits making a definite absorption correction (plane parallel layer) for the intensity measurements of the diffracted X-rays, and the diffraction pattern of the cell walls does not interfere with the liquid pattern itself. The absorption and the scattering of the cell walls are small compared to the corresponding values for the liquid. Monochromatic silver $K\alpha$ radiation reflected from a rock salt crystal was used in a vacuum camera, thus eliminating air scattering and continuous background.

The diffraction pattern, recorded photographically, shows three distinctly visible rings and a fourth ring is revealed upon closer inspection (Fig. 1).

Figure 2 shows the experimentally observed intensity (I_{exp}) properly corrected together with the coherent (I_{coh}) and incoherent (I_{inc}) scattering to be expected theoretically matched at $(\sin\theta)/\lambda$ of about 0·7 for quantitative analysis.

From a Fourier analysis the atomic distribution curve in the liquid has been obtained and is shown in Fig. 2B. The first neighbours are to be found at 3·8 Å (3·82 Å in the solid). The area under this peak gives the number of the first neighbours. From the various patterns which we have obtained this number was determined as 9·6, 10·1 and 10·3 nearest neighbours.[2] Further concentrations corresponding to the outer neighbours occur at 5·4, 6·7 and 7·9 Å. The number of neighbours and their distances, as they occur in the solid (Fig. 2B), indicate how far the liquid structure retains the arrangement found in the solid.

The results show that the type of co-ordination, as found in the solid, is retained in the liquid, but the transition solid–liquid in argon produces a disorder already pronounced in the reduction of number of first neighbours, as compared to $N_1 = 12$ in the solid.

Theoretical investigations[3] indicate that such a reduction in N_1 is necessary to account for the thermodynamical data. Recent investigations by O. K. Rice[3] lead to 10–10·5 nearest neighbours for argon in the temperature range of our experiments, in good agreement with the results obtained.

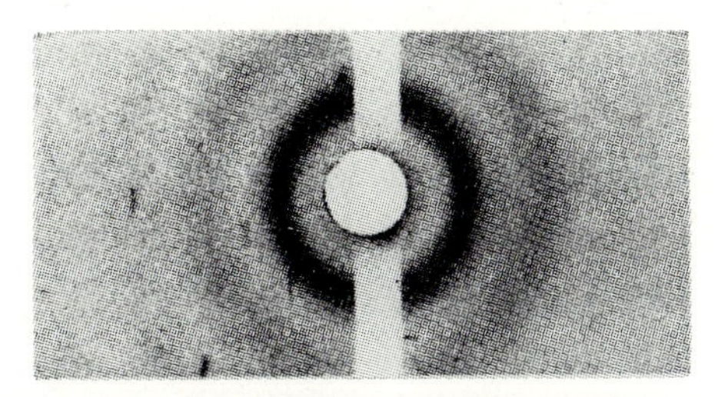

FIG. 1. Diffraction pattern of liquid argon showing diffraction rings at $(\sin\theta)\lambda = 0{\cdot}152, 0{\cdot}284, 0{\cdot}410$ and a faint ring is indicated at $0{\cdot}59$. The Laue spots are due to the mica windows.

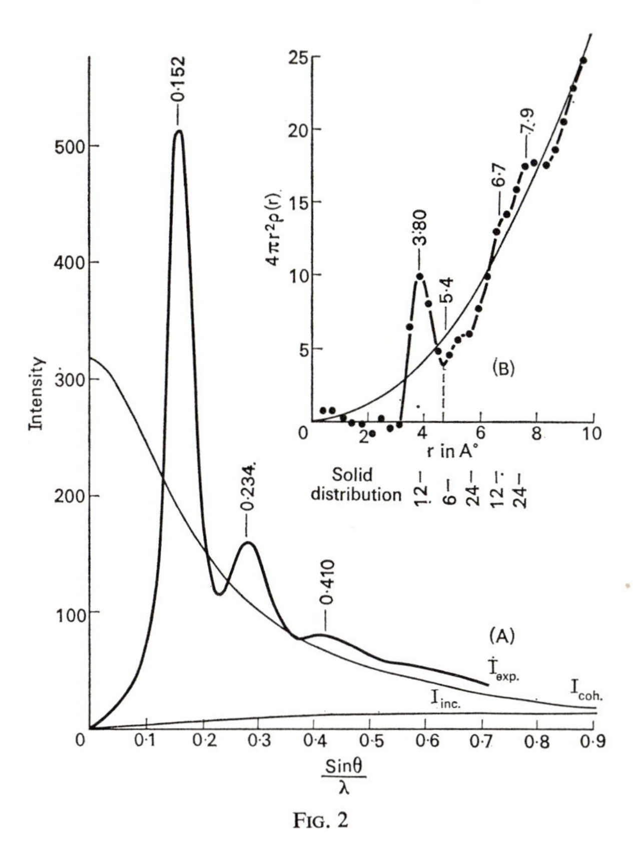

FIG. 2

The substance of this work was reported at the Pittsburgh meeting of the American Physical Society during June 20–22, 1940.

NOTES

[1] R. H. Fowler and E. A. Guggenheim, *Statistical Thermodynamics*, § 808, and following.

[2] A. Eisenstein and N. S. Gingrich (*Bull. Amer. Phys. Soc.* **15,** No. 2; April 1940) in a preliminary report give 7 atoms at 3·9 Å as nearest neighbours. This result, which seems rather difficult to reconcile with the theoretical

predictions, was obtained with an experimental arrangement different from ours and in a different p–T range. It might be pointed out that at higher molal volume the curves of Rice indicate a possible change in co-ordination, but scarcely so low as the preliminary results of these authors indicate.

[3] Fowler and Guggenheim, *loc. cit.*, and O. K. Rice, *J. Chem. Phys.* **7**, 136 and 883 (1939).

5

Structure of the Wood Used in Violins†

THE many recent investigations on string instruments[1] deal mainly with two problems: the theory of the mechanical and acoustical behaviour of the different parts of the instrument (and its experimental verification), and the analysis of the tones produced by the instruments. The question of the proper choice of material has made scarcely any progress since the fundamental investigations of F. Savart.[2] Since it has been repeatedly stated that age, treatment and varnish change the character of the wood, we investigated the structure of the wood in violins of different origin[3] with X-rays. Copper $K\alpha$ (in a few cases also molybdenum $K\alpha$) rays fall (*a*) through the *F*-hole on to the back of the instrument or (*b*) are reflected from the edges of either top or back.

In all of the instruments investigated, we found that the spruce used for the top shows definite fibre structure, giving almost identical patterns (Fig. 1*a*; for comparison see Fig. 1*b*). But the patterns from the wood used for the back (mostly maple) are different for instruments of different tone quality. Instruments with an even and smooth tone quality, especially for higher pitch (*E*-string), show an almost complete lack of orientation in the wood used for the back (Fig. 2). The maple used in instruments which have a harsh tone quality in general, weak response and shrill upper register show marked fibre structure (Fig. 2*b*). Since we found that instruments two hundred years old may show such a pattern, it is clear that the ageing of the wood after cutting and working does not change its structure. Whether a special treatment of the wood or a special varnish has been used by the

† *Nature* **134,** 23 (1934), with W. I. Caldwell, and *ibid.* **137,** 663 (1936).

Italian makers is so far not certain; we found only in one case a diffraction pattern containing one ring which could not be interpreted as belonging to cellulose.

Investigation of untreated maple as used for violin making has shown that sometimes, although rarely, such maple will show as small an amount of orientation as the wood found in good violins. Occasionally modern violins with properly chosen wood for which Italian varnish and treatment of the wood are not used, show an evenness in tone comparable with the old instruments. All this seems to indicate that the proper selection of the wood is more important for the quality of the instrument than treatment and varnish. We found several instruments with the proper wood, but a poor tone quality. This, of course, can be due to the faulty model of the instrument, but in two cases investigated, a radiographic X-ray study of the interior of the violin revealed a great number of crude repairs which necessarily would impair the tone of the instrument. We have found that these radiographic studies are of great value in supplementing the knowledge of the connoisseur and collector.

Our investigation indicates that for a fine instrument only the top should be characterised by different velocity of sound in different directions, whereas the velocity of sound in the back should be the same in all directions so as to produce the best results.

A Simple Method for Testing Homogeneity of Wood

Some time ago† we found that there seems to be a characteristic difference in the structure of the wood used for the building of string instruments: X-ray investigations have shown that the top always exhibits a very marked fibre structure, whereas the back in instruments of good tone quality is nearly homogeneous.

The question arises whether it is possible to find these differences by methods which might have been available to the Italian makers of the classical period? I found that it is possible to obtain this

† *Nature* **134,** 23 (1934).

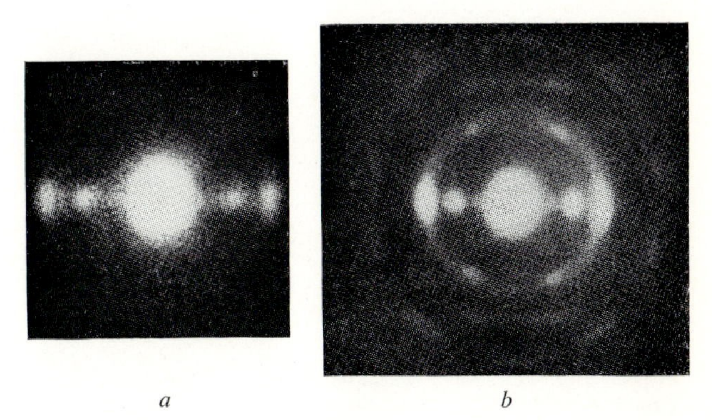

a *b*

FIG. 1. (*a*) Spruce (top) from "Geneva" Guarnerius; (*b*) ordinary spruce.

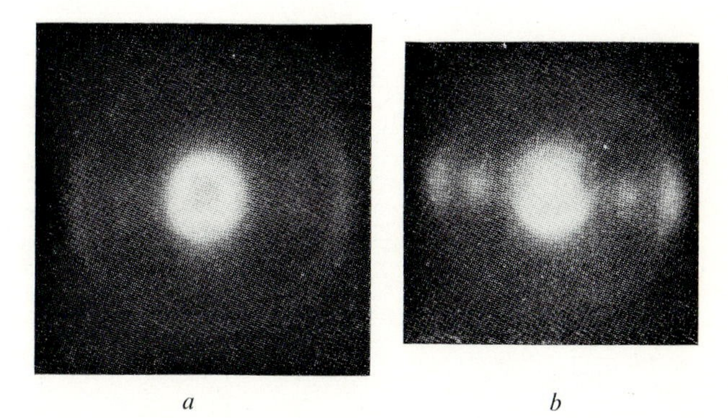

a *b*

FIG. 2. (*a*) Maple (back), J. Guarnerius; (*b*) maple (back) of modern violin.

information by using heat conductivity in the different directions of the wood as an indicator of its homogeneity. It is well known that a thin layer of wax applied to a crystal face will melt into a figure of definite contour (isotherm) if the crystal is touched at one point with a hot wire. The same method can be easily applied to wood, and one finds that the isotherm on a piece of wood cut vertical to the fibre is always a circle, except where a knot produces an inhomogeneous region. The isotherm on a cut parallel to the grain varies in its outline for different materials. The ratio of the axes for pine used for the top of violins has been found as high as 1·95, and for nearly homogeneous maple used for the back 1·15. We have obtained recently, through the courtesy of Dr. A. Koehler, director of the U.S.A. Forest Products Laboratory, Wisconsin, some samples of white ash which range, as revealed by X-ray investigations, from very marked fibre structure to almost complete homogeneity. The same variation and exactly the same order has been found by using the isotherm method.

It is possible that such a method, discovered accidentally, may have been used by the instrument makers, since many of the old instruments exhibit branding marks even if the maker did not use a brand for the identification of his instruments.

Notes

[1] C. V. Raman (summary of all of his important papers), *Handbuch der Physik*, vol. 8, pp. 355–424; H. Backhaus, *Handbuch der Experimentalphysik*, vol. 17/3, pp. 177–256; 1934. R. B. Abbott, *Phys. Rev.*, August 1933; March 1934.

[2] F. Savart, *Mémoire sur la construction des instruments à cordes et à archet*, 1819. The much more important investigations published in *l'Institut*, 8, 55, 69–70, 91, 122, 1840 seem to be entirely forgotten. We found them only quoted in E. Heron-Allen's book on *Violins and Violin Making* and have seen them myself only now. Savart's conclusions should certainly be checked with more modern acoustical methods.

[3] We are indebted to Messrs. Lyon and Healey and Wm. Lewis and Son, Chicago, for the possibility of investigating instruments of the following makers: A. and H. Amati, Stradivarius, J. Guarnerius, J. B. Guadagnini, C. Bergonzi, M. Bergonzi, Montagnana, Storioni, Vuillaume, Pique and several modern makers, altogether 24 instruments.

6

Electron Diffraction Patterns†

The Origin of the "Extra Rings" in Electron Diffraction Patterns[1]

The diffraction patterns produced by electrons differ from the diffraction pattern produced by X-rays not only in the intensity distribution but also in the appearance of new lines.

These can be produced (a) by surface layers, since cross gratings may also produce sharp rings and (b) by the refraction of electrons on "electron smooth" surfaces. If the electron beam enters and leaves at the same crystal surface the Bragg equation $\sin \vartheta = n\lambda/2d$ has to be replaced by $\sin \vartheta' = n\lambda/2d(1 - 4d^2V_0/150n^2)^{\frac{1}{2}}$ [2] and this may lead also to the appearance of lines not existing in the X-ray diffraction pattern. For low velocity electrons it has been found[3] that between the ordinary interference maxima there appear new maxima at a distance halfway between the regular maxima (so-called "fractional orders"). It has been assumed that these are due to an adsorbed gas layer (e.g., hydrogen).[4]

In the case of fast electrons, rings have been observed which are entirely inside the ordinary diffraction pattern and which also have been interpreted as fractional orders.[5] There is no theoretical reasoning to lead to the existence of such fractional orders.

During the last two years we have been using a new method[6] for making thin films for electron diffraction. The substance to be investigated is deposited in a high vacuum upon a volatile material which is later pumped off, leaving the free film. For most

† *Phys. Rev.* **48,** 101 (1935) and *Proc. Amer. Phil. Soc.* **76,** 766 (1935), with H. J. Yearian and J. D. Howe.

experiments purest naphthalene has been used. The following substances have been investigated: aluminum, cadmium, copper, cuprous oxide, gold, lead, lead oxide, nickel, silicon and silver. For all these substances we have found electron diffraction patterns containing sharp "inner rings" in positions similar to those reported by Rupp for gold. It is possible to interpret these rings as fractional orders of the ordinary interference maxima. Such an interpretation would necessarily lead to the expectation that the values of sin ϑ/λ are different for different substances. Table 1 gives a summary of the values of sin ϑ/λ as observed in our experiments. For comparison we have added the values from Rupp's work on gold (Au(R) in table).

The values of sin ϑ/λ are identical for the different substances indicating that the diffraction pattern is not due to the substance itself but to an adsorbed surface layer. Since our films were produced by depositing on naphthalene we suspected that a thin layer of naphthalene might be responsible for the diffraction pattern observed.[7] An X-ray powder pattern of the naphthalene used in our experiments gave values for sin ϑ/λ, most of which are identical with the ones of the extra inner electron diffraction rings. The similar values which Rupp has found for sin ϑ/λ in the absence of naphthalene are apparently due to adsorbed layers of organic substances which have similar spacing due to the similar C–C distances. Rupp states expressly that his rings cannot be due to traces of grease vapor since at the high voltages used (200 kV) the vacuum would break down in the presence of vapor. We have found that the adsorbed layers cannot be driven off by heating the film up to 500°C and also withstand electron bombardment.

The fact that the rings appear extremely sharp leads to the conclusion that we are dealing not only with thin crystal layers but with random orientated cross gratings. Also in this case, as has been shown by Laue,[8] sharp diffraction patterns are observed. This conclusion is further supported by the intensity distribution which shows characteristic differences from the distribution in space lattices.[9] The intensity sets in sharply at a certain angle and falls off from there. In some of our diffraction patterns we observe a blackening starting from one ring and filling continuously the interval on the plate to the next ring.

Recently we have been able to obtain thin films for electron diffraction by depositing on ammonium chloride which is later driven off by heating to 200°C in a vacuum. In this case no "inner rings" have been observed.

That these rings can be produced by contact with organic vapors (grease, oil) we have shown in the following manner: Two gold films have been made under exactly the same conditions. One film was exposed to organic vapors (from a grease joint), the other one was held in a perfect vacuum. The first film gave "inner rings", the second not. After exposing the second film to grease vapor the "inner rings" have been observed too. Our experiments seem to indicate that even the presence of small naphthalene nuclei, so sparsely distributed that they themselves will not give a diffraction pattern, will induce the deposition of other organic matter which will then produce cross gratings responsible for the "inner rings" observed.

Summarizing we conclude that the so-called fractional orders, as observed in the diffraction of fast electrons, are due to adsorbed layers of organic substances, forming thin crystalline layers and random orientated cross gratings.

Intensity Distribution in Electron Diffraction Patterns

ABSTRACT

A universal camera has been constructed for the production of electron diffraction patterns of single crystals, powders, liquids and vapors. It is possible in this camera to take diffraction patterns from a surface orientated at the incoming beam in different directions, as well as diffraction patterns produced by penetration of the material. Investigations can be carried out from liquid air temperature up to 400° and the changes in structure can be recorded, without changing the vacuum, by an automatic mechanism allowing the succession of 49 exposures with an exposure time which can be preselected and automatically provided.[10]

The material was used mostly in the form of thin films which have been prepared with a new method which allows the production of unsupported films only a few atoms thick.[11] Using this method it has been possible to find the origin of "extra rings" in electron diffraction patterns as due to surface layers.[12]

Electron diffraction patterns have been obtained with voltages ranging from 80 kV down to 15 kV. The results show that the behavior at high voltages is entirely different from that at low voltages.

Atom factor determinations for copper in pure copper and in cuprous oxide, for zinc, cadmium, gold, silver and palladium[13] have shown that the wave mechanical formula

$$F_{el}(\vartheta) = \frac{Z}{2E} \frac{1 - F_x}{\sin^2 \vartheta/2} = \frac{1}{\sqrt{E}} \Sigma(2l + 1) P_l(\vartheta) \delta_l$$

as given by Mott is fulfilled at the higher voltages. At the lower voltages one would expect that this formula above should be replaced by the better approximation as given by Massey and Henneberg

$$F_{el}(\vartheta) = \frac{1}{2i\sqrt{E}} \sum_l (2l + 1) P_l(\cos \vartheta) (e^{2i\delta_l} - 1).$$

The experiments show[14] that while the behavior in general is similar to the one indicated by this formula, the deviations are so great as to indicate an entirely different phenomenon not described by the theory. By using materials of different structures and varying in atomic number by one it was possible to obtain a scattering curve covering more or less completely the whole region of angles investigated (up to 10°). The results indicate that it is necessary to extend the theory in two directions: better approximation of the atomic field, and interaction between lattice and electron wave including the influence of surface or cross lattices.

We reported in February 1935 on the completion of a Universal type of electron diffraction camera. The camera has been described since then in the *Review of Scientific Instruments* (Vol. 7, 26, January 1936). This apparatus allows investigation of electron scattering by substances in any desired state of aggregation: single crystals, solid powders, liquids or gases. It also allows intensity measurements by exposures taken in a definite ratio which can automatically be pre-selected.

The whole operation of the camera: changing plates and exposing them for any desired length of time, selecting of voltages—is all done automatically from a universal switch board.

The method of making thin films for the diffraction by solids has been extended by applying ammonium chloride as a new inorganic base for condensation of material.

In this way it was possible to show that the extra rings, as observed in electron diffraction, are actually due to organic layers formed on the surface of the material. If any such organic layer is excluded by producing the films on ammonium chloride and avoiding exposure to organic vapor, inner rings are never observed. If, on the other hand, such films are exposed to organic vapors, formation of the inner rings has been found. It seems, therefore, that the troublesome question in electron diffraction experiments regarding the existence of so-called forbidden rings is now cleared up and solved by proving that these rings are due to organic impurities.

The method of making thin films was applied to the investigation of the scattering of electrons of varying energy. It has been pointed out before that electron diffraction allows observation of nucleus and surrounding electron cloud. If electrons of low energy are scattered by an atom of high atomic number, the distortion of the incoming electron wave in the potential field of the nucleus has to be taken into account. This means that the formula for scattering usually applied and leading to

$$F(\vartheta) = \frac{e^2}{4E} \frac{(Z - f)}{\sin^2 \vartheta/2}$$

cannot be used anymore, but one has to use the exact solution of the wave equation (written in Hartree units)

$$(\Delta + 2V + E)\psi = 0,$$

which leads to the following expression:

$$F(\vartheta) = \frac{-i}{2k} \Sigma(2l + 1) P_l(\cos \vartheta) (e^{2i\delta_l} - 1).$$

We have pointed out in a former report that our results for the intensity of electron scattering by gold are markedly different from the ones obtained at much higher velocities by Rupp, and that we think this is due to an influence of the potential of the atoms.

In discussing these results with Dr. H. A. Bethe, he pointed out to us that probably these deviations should become still more pronounced if we would choose lower voltages.

The actual limit for voltage is given by the relative transparency of the different films and the necessary time for exposure. We found that electron velocities to about 20 kV can easily be used. Below 20 kV the sensitivity of the photographic plate decreases suddenly and exposure time becomes much longer. We have been able, however, to produce diffraction patterns at about 15 kV. As the first materials to be tried we have used gold and silver. The *K* excitation limit of gold is about 80 kV; the *K* excitation limit of silver is about 25 kV. It is therefore easily possible to obtain diffraction patterns for gold films at electron energies well below the *K* excitation limit. This is also possible in the case of silver, and we have actually found much stronger deviations in about the same direction as the ones which we had observed originally with gold at 80 kV.

In the case of solid crystals the atom factor measures the scattering from the electrons of the different atoms in the crystal, and in this way it is possible to check the theoretical prediction. Since crystals reflect only under certain Bragg angles, it is impossible to obtain a complete scattering curve, but only points on this curve can be obtained in the different regions. Therefore, we have made use of the following idea: if the electron scattering is actually influenced by the potential field of the atom only, then one would not expect a great deal of change if we compare the scattering power of atoms which are not more than one unit in atomic number apart. As such atoms we have chosen Pd 46, Ag 47, and Cd 48. In this way we obtain scattering under different angles since the different elements crystallize in different systems (Cd hexagonal, Ag, Pd cubic, but with different lattice constant), and therefore a more complete scattering curve has been obtained.

The scattering curves have been measured for electron velocities from 80 kV down to 17 kV. At 80 kV, at a value far above the *K* excitation limit of the elements, the ordinary scattering formula represents the results adequately. In the lower voltages, strong deviations are observed which exhibit a marked periodicity in a region of not more than 10° of scattering angle.

Qualitatively this result seems to agree with the results obtained for the scattering of slow electrons (below 1 kV); also in this case a periodic change of scattered intensity under different angles has been observed. This variation, however, extends over a wide range of angles and not over as small a range as we have mentioned above.

It seems, therefore, that we are dealing here with a new phenomenon which has not been observed before. This conclusion was confirmed by calculations of the theoretical scattering of 17 kV electrons from gold and silver, carried out under the direction of Dr. Nordheim by Miss Fry. This calculation shows that also the exact solution of the wave equation, as given above, is not able to explain this effect.

One might consider first that this is due to the use of a Thomas–Fermi field for the atom, and that this should be properly replaced by a Hartree field. One can see, however, that since we are dealing with scattering at small angles, large values of l have to be considered (in our notation of the phase angle δ_l) and therefore, if such fluctuations occur one would expect that they should occur also in the phases δ_l themselves. Since the Born approximation formula can be written in the following form also,

$$F(\vartheta) = \frac{i}{k} \Sigma(2l + 1) P_l(\cos \vartheta)\delta_l,$$

one would expect that such large fluctuations should occur in the phases themselves if using a Hartree field and that it would make a difference even for the approximate theory whether one uses a Thomas–Fermi or a Hartree field. We have plotted scattering curves using both fields wherever Hartree calculations are available, but no such fluctuations have been observed, and the curves are perfectly smooth.

We came, therefore, to the conclusion that the effect which we have observed can hardly be explained by a distortion of the electron wave by the free atom alone, but must be due also to the field of the atom in the crystal. It seems to be a type of dynamic scattering of electrons which does not produce new lines but a definite intensity change of the ordinarily observed Bragg reflections.

This hypothesis is being tested now by carrying out experiments on the scattering of electrons from molecular vapors using the same range of velocity as we have applied for the investigation of crystals. It is not possible to use scattering from free atoms in a monatomic vapor since for these small angles the ratio of inelastic to elastic scattering is so great as to even out the changes in relative scattering intensity under different angles for the elastic collisions.

Summarizing, we can say that it has been shown that while the kinematic theory of electron diffraction using a statistical distribution of electrons in the scattering atom fits the experimental data for high energy electrons, this theory has to be replaced by a dynamic theory for electron energies small as compared with the K excitation limit of the scattering element.

Notes

[1] Grateful acknowledgement is made to the American Philosophical Society for a grant-in-aid of this research.

[2] K. Lark-Horovitz and H. J. Yearian, *Phys. Rev.* **43,** 376 (1933).

[3] L. H. Germer, *Zeits. f. Physik* **54,** 408 (1929); E. Rupp, *Ann. d. Physik* **5,** 453 (1930).

[4] This interpretation, however, is difficult to understand because of the strong intensity of the fractional orders.

[5] E. Rupp, *Ann. d. Physik* **10,** 933 (1931).

[6] K. Lark-Horovitz, H. J. Yearian and E. M. Purcell, *Phys. Rev.* **45,** 123 (1934); J. D. Howe and E. M. Purcell, *Phys. Rev.* **47,** 329 (1935).

[7] The experiments with Au, Ag, Cu, Cu_2O, Cd, Ni deposited on naphthalene and interpreted as a bove have been reported at the Pittsburgh meeting of the Am. Phys. Soc.

[8] M. Laue, *Zeits. f. Krist.* **82,** 127 (1932).

[9] A. Steinheil, *Zeits. f. Physik* **89,** 50 (1934).

[10] H. J. Yearian and J. D. Howe, *Rev. Sci. Inst.* **7,** 26–30 (1936).

[11] K. Lark-Horovitz, J. D. Howe and E. M. Purcell, *Rev. Sci. Inst.* **6,** 401 (1935).

[12] K. Lark-Horovitz, H. J. Yearian and J. D. Howe, *Phys. Rev.* **48,** 101 (1935).

[13] K. Lark-Horovitz, H. J. Yearian and J. D. Howe, *Phys. Rev.* **47,** 331 (1935).

[14] H. J. Yearian and J. D. Howe, *Phys. Rev.* **48,** 381 (1935).

7

Fission of Uranium†

Fission Tracks on the Photographic Plate

Photographic plates (Eastman fine-grain α, Ilford R_2) bathed in alcoholic solutions of uranium nitrate and thorium nitrate have been exposed to slow and fast neutrons and the resulting tracks have been measured microscopically with a filar micrometer. With fast neutrons new types of tracks besides the ordinary alpha-particle tracks have been observed on both the U and Th treated plates; with slow neutrons these tracks have been found only on the U plate. We conclude that these tracks are due to fission fragments because: (a) Their range is different from the range of the U α-particles. (b) The reciprocal grain spacing decreases toward both ends of the track, in contrast to the appearance of α-tracks. (c) Distribution curves of the observed grain spacings show a minimum distance far more frequently than with α-tracks. (d) Recoil protons issuing from the track are frequently found; and from stereoscopic projection their energy can be estimated and related to the energy of the fission product at this point of the path in good agreement with the calculation of Knipp and Teller.[1] Most of the tracks are asymmetric but some symmetric tracks have been observed. To study fission processes due to cosmic radiation, U and Th treated plates as well as plates treated with thallium iodide solution have been sent into the stratosphere reaching a height of about 30 km. Fission tracks like the ones observed in the laboratory were obtained only on the U plate but not on the Th plate

† *Phys. Rev.* **59**, 941 (1941), with W. A. Miller, and *ibid.* **60**, 156 (1941), with R. E. Schreiber.

indicating that most of the neutrons in the cosmic radiation even at this altitude are too slow to produce the fission.

Uranium Fission with Li–D Neutrons: Energy Distribution of the Fission Fragments

The energy distribution of the fission fragments from slow neutron bombardment has been reported by different observers (Jentschke and Prankl,[2,3] Haxel[4]). Two energy groups are found corresponding approximately to 60 and 95 MeV. These groups are separated by a valley about one-fourth the height of the peaks if the number of particles is plotted against energy. Some observers have also used fast neutrons (Ra–Be neutrons; [1]D–D neutrons, Kanner and Barschall[5]) and in this case the valley is considerably raised. One can, therefore, conclude that with fast neutrons symmetrical fission should be observed, and we started some time ago to investigate the energy distribution of the fission fragments with Li–D neutrons.[6]

The recoil energies were observed in an ionization chamber with a linear amplifier and recording oscillograph which was calibrated with alpha-particles and artificial pulses to extrapolate to the energy range used. Two different uranium samples were used: (a) uranium metal of 0·18 mg/cm² sputtered on an aluminum foil of 0·17 mg/cm², which was used both for measurements of total ionization and of single fragment energies, and (b) uranium oxide deposited electrolytically on a brass disk.

By surrounding the ionization chamber with 25 cm of paraffin and water, the fission with slow neutrons has been recorded and the results found in agreement with those of former investigators. Both types of targets show two peaks (with single particles) at 64 and 95 MeV.

The paraffin was then removed and the slow neutrons filtered out with cadmium shields. We now observed new modes of fission. Besides the two peaks observed before, there now appeared a peak corresponding to 86 MeV. This peak we assigned to the symmetrical fission. There also appeared peaks at 52 and 110 MeV,[7] which apparently correspond to highly asymmetrical fission such as has been found by Nishina[8] who identified Hg and Bi among fission fragments from Li–D neutron bombardment.

About 5% of the total number of tracks give a peak of much higher energy, at 130 to 135 MeV, which we believe is due to a new type of fission as predicted by Present:[9] ternary or triple fission. Some fragments have even higher energy, 145 MeV, but are quite rare; they are 0·5% of the total number of tracks.[10]

By using the thin foil and observing the total ionization, these results are substantiated. With slow neutrons the main peak occurs at 164 MeV. With fast neutrons the main peak appears at 172 MeV and there is a definite indication of fragments of a still higher energy corresponding to 190 to 200 MeV.

The measurement of the energy distribution thus confirms the process of symmetrical fission corresponding to the chemical evidence in the experiments of Yasaki[11] and Segrè and Seaborg[12] and agrees also with the observation of the fission tracks on the photographic plate as described recently by Lark-Horovitz and Miller.[13]

The existence of high energy fragments in appreciable numbers is attributed to ternary fission. The masses of the fragments lie in the same region as those from ordinary asymmetric fission, therefore no chemical evidence is available. From Present's calculations,[9] about 20 MeV more energy should be released in this process. This is in agreement with our observations. It is not possible, however, to distinguish between triple fission and a binary fission with the fragment masses in the ratio 2 : 1 from ionization-chamber measurements alone. It will be necessary to use a cloud chamber or direct photographic detection. This work is now in progress.

Notes

[1] J. Knipp and E. Teller, *Phys. Rev.* (to be published).
[2] W. Jentschke and F. Prankl, *Akad. Wiss. Wien* **148,** 372 (1939).
[3] W. Jentschke and F. Prankl, *Physik. Zeits.* **40,** 706 (1939).
[4] O. Haxel, *Zeits. f. Physik* **112,** 681 (1939).

[5] M. H. Kanner and H. H. Barschall, *Phys. Rev.* **57,** 372 (1940).

[6] This conclusion has been confirmed since this work was begun by the discovery of fission products in the region between Pd and In by Yasaki and Segrè and Seaborg.

[7] Pulses as high as 110 MeV have been observed by Kanner and Barschall.

[8] Y. Nishina *et al., Nature* **144,** 547 (1939).

[9] R. D. Present, *Phys. Rev.* **59,** 467 (1941).

[10] Jentschke and Prankl record a few fragments of 140-MeV energy and attribute them to the total energy of both fragments even in a thick target experiment. This explanation is excluded by our experimental arrangement.

[11] T. Yasaki, *Inst. Phys. Chem. Research Tokyo* **37,** 457 (1940).

[12] E. Segrè and G. T. Seaborg, *Phys. Rev.* **59,** 212 (1941).

[13] K. Lark-Horovitz and W. A. Miller, *Phys. Rev.* **59,** 941 (1941).

8

Radioactive Indicators, Enteric Coatings and Intestinal Absorption†

CERTAIN medicaments cannot be administered orally without either irritating the stomach tissues or becoming destroyed by the gastric juice. To avoid these effects they are given in pills with enteric coatings. Such coatings are supposed to pass the stomach undissolved and to dissolve only in the small intestines.

Since the introduction of enteric coatings by Unna in 1884, various methods have been tried to test their efficacy. In recent years X-ray radiology has been used.[1] The methods used so far do not detect cracks in the coating, which may lead to a leakage and consequently to destruction of the contents by the gastric juice.

Radioactive materials as fillings are obviously admirably suited for such tests. Any crack in the coating is detected at once by the appearance of the radioactive material in the blood-stream as measured by the activity in the hand, and the capsule itself can be easily followed on its way through the digestive tract until it is dissolved. The experimental arrangement, therefore, is the following: Ordinary gelatin capsules are filled with about 0·3 g of radioactive sodium chloride obtained by bombardment with 8·5 MeV deuterons for about 2–6 μA-hr. The initial activity of the sample is of the order of 500–700 μC. The capsule is coated with a mixture of shellac, castor-oil and alcohol following a formula of Prof. C. O. Lee of the School of Pharmacy, Purdue University. The capsule is administered orally and is followed through the digestive tract with the aid of a movable bell-type counter.

† *Nature* **147,** 580 (1941), with H. R. Leng.

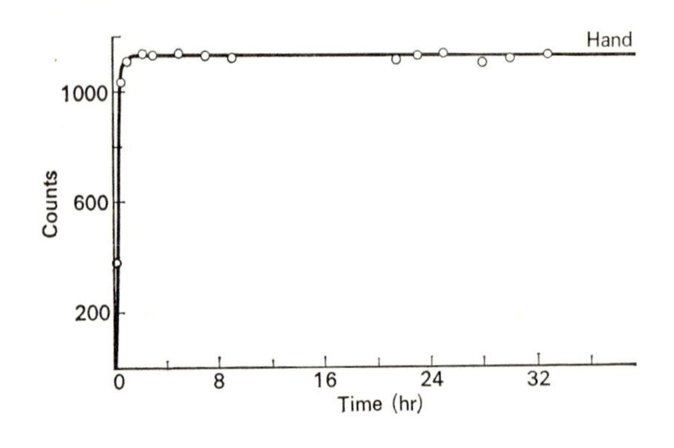
Hand
1000
600
200
Counts
0
8
16
24
32
Time (hr)

FIG. 1

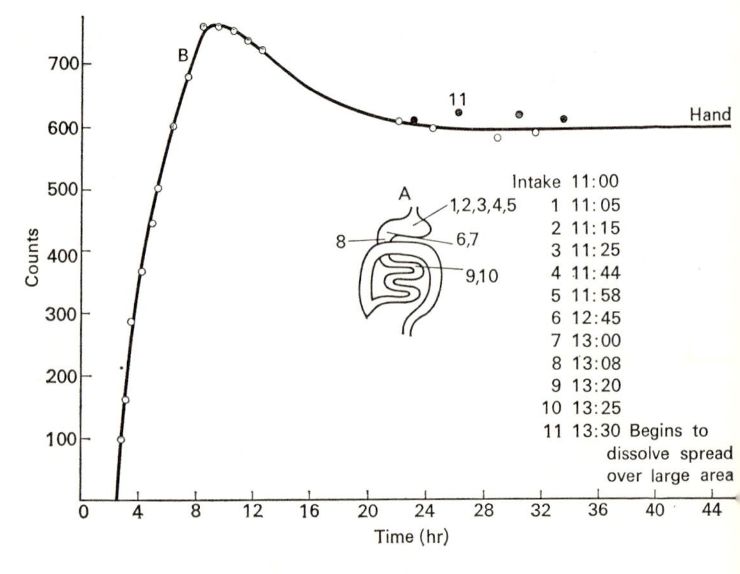
700
600
500
400
300
200
100
Counts
B
11
Hand
A
1,2,3,4,5
8
6,7
9,10
Intake 11:00
1 11:05
2 11:15
3 11:25
4 11:44
5 11:58
6 12:45
7 13:00
8 13:08
9 13:20
10 13:25
11 13:30 Begins to dissolve spread over large area
0
4
8
12
16
20
24
28
32
36
40
44
Time (hr)

FIG. 2

Simultaneously, the activity of one hand of the patient is measured to determine the movement when the active material has reached the blood-stream.

Figure 1 shows the activity as a function of time when the salt has been administered orally without the coating. The first sign of activity in the hand is detected a few minutes after intake; it then increases steadily to reach a constant value in about 3 hours.[2]

Figure 2A shows the location of a coated pill in the digestive tract, throughout the experiment. In this particular case, it is located for about one hour in one position in the stomach; then it moves towards the pylorus and can be located again in the small intestines. After 2½ hr the movable counter indicates that the pill dissolves and the activity is spread over a large area. Now the activity appears also in the hand, which was, up to this time, entirely inactive.

Figure 2B shows the activity in the hand throughout the experiment. The first activity is observed after about 2½ hr, when the pill starts to dissolve; it reaches a maximum after about 8 hr (5½ hr after it started to dissolve) and then decreases to equilibrium value.

Figure 3 was obtained with a pill which apparently was leaking, for an activity in the hand can be detected half an hour after intake. The pill itself could be followed for about 8 hr, when it finally dissolved in the intestines. At this time the activity in the hand is raised, reaches a maximum in about 1 hr and falls again to a level higher than the first one, since now the total activity is distributed throughout the body.

Besides the handcoated pills, commercial coatings have been tested. In this case some 100 or 1000 pills are coated at the same time, and the few active samples can be easily segregated with the aid of a counter.

The results in Table 1 indicate that the majority of the handmade pills dissolve in the small intestines.

In connexion with other experiments on the distribution of the sodium, potassium and chlorine ions in the body, we want to direct attention to the fact that the curve for the time-rate of

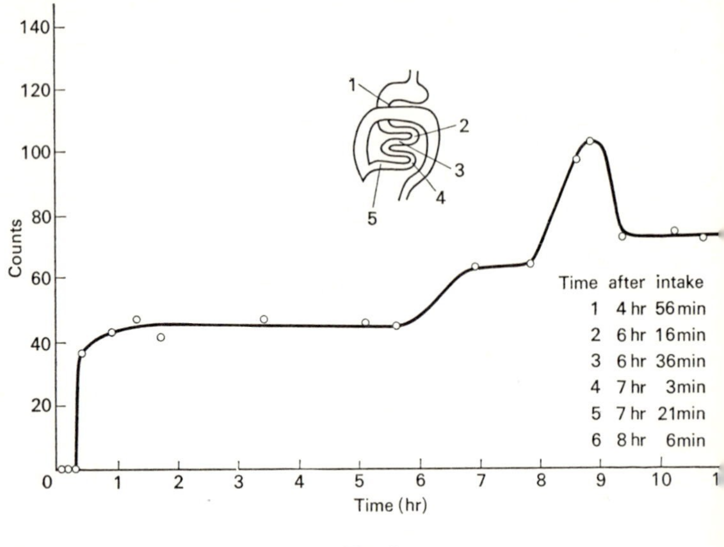

FIG. 3

absorption from the stomach is entirely different from the absorption-curve from the small intestines. In the first case the absorption reaches a level which remains constant for hours and decreases only after a large amount of the salt has been excreted. The absorption from the intestines raises the salt level temporarily and equilibrium is reached through back-diffusion, indicating

TABLE 1

	Total no. of exp.	Dissolved in intestines	Dissolved in stomach	Unknown	Not dissolved
Handcoated pills	21	14 (66%)	2 (10%)	3 (14%)	2 (10%)
Commercial coatings	10	5 (50%)	4 (40%)	1 (10%)	

that a local surplus produced by a sudden release of salt causes a real absorption through the wall of the intestines which is adjusted later by back diffusion and ionic exchange.

A detailed description of these experiments will be given elsewhere.

References

1. J. T. Corley and C. O. Lee, *J. Amer. Pharm. Assoc.* **27** (5), 379–384 (1938).
2. J. G. Hamilton, *Proc. Nat. Acad. Sci.* **23,** 521 (1937).

9

Resistivity and Thermoelectric Power of Germanium Alloys (Abstracts of first post-war presentation of war-time research)†

Electrical Properties of Germanium Alloys. I. Electrical Conductivity and Hall Effect

Germanium samples purified by high vacuum treatment have been alloyed by adding from 0·001 to 1·0% of metallic impurities. The electrical conductivity and transverse Hall effect of these samples have been investigated over temperatures ranging from $-180°C$ to $650°C$. Plotting log ρ (resistivity) and log R (Hall constant) against $1/T$ shows that the resistivity at low temperatures decreases with increasing temperature, increases around room temperature and then drops sharply with a slope identical for the various samples. The Hall curves indicate electron (*N* type) or hole (*P* type) conduction, depending upon the type of impurity. *P* type samples show reversal of Hall effect, and the slope at high temperatures is identical for all samples, *P* and *N* alike, indicating that germanium behaves at low temperatures as an impurity semiconductor, but is at high temperatures an intrinsic semiconductor with energy level separation of about 0·76 volt. Hall values show that the number of current carriers ranges from about 10^{15} up to 10^{19} per cc. The temperature behavior of mobility, determined by R/ρ, cannot be explained as due to lattice scattering alone, but indicates the existence of another scattering mechanism, especially at low temperatures.

† *Phys. Rev.* **69**, 258, 259 (1946), with V. A. Johnson, A. E. Middleton, E. P. Miller, W. W. Scanlon, and I. Walerstein.

Theory of Resistivity in Germanium Alloys

The temperature dependence of resistivity of germanium alloys is considered in three ranges: (1) the impurity range of low temperatures with conduction due to impurity electrons or holes, (2) the transition range with contributions from both impurity and intrinsic electrons and holes, and (3) the intrinsic range of temperatures so high that the numbers of electrons and holes are equal. Resistivity in the impurity range is the sum of resistivity due to lattice scattering (proportional to $T^{\frac{3}{2}}$) and resistivity due to impurity scattering. Mean-free-path calculations indicate that the impurity mean-free-path increases and the lattice mean-free-path decreases with rising temperature. At a given temperature the lattice mean-free-path is the same for all germanium samples, but the impurity mean-free-path varies widely with impurity content. In the transition and intrinsic ranges conductivity is the sum of conductivities due to electrons and holes; their values are based upon Hall constant values. Combination of these two methods allows complete synthesis of the experimental results throughout the entire temperature range.

Electrical Properties of Germanium Alloys. II. Thermoelectric Power

The thermoelectric power, Q, of germanium alloyed with various metallic impurities has been studied as a function of temperature from −180°C to 650°C. For this purpose thermocouples have been imbedded into the body of the semiconductors and temperature gradients of 3 to 30°C/cm have been used depending on the temperature range and the size of the sample. The thermoelectric behavior parallels R, the Hall effect. R and Q are positive for P type samples at low temperatures, reversing sign at high temperatures. The thermoelectric power, Q, becomes zero at a temperature near but not identical with the temperature where R becomes zero. In most cases Q as a function of temperature rises in the low temperature range, passes through a maximum and decreases again near room temperature. The position and height of the maximum depends on the amount and type of impurity used. The temperature dependency of Q for samples containing the same type of impurity in varying amounts are represented by a family of similar curves: with increasing impurity content the position of the maximum shifts to higher temperatures and lower values.

Theory of Thermoelectric Power in Germanium

Thermoelectric power Q in semiconductors has been calculated under conditions of simultaneous conduction by positive and negative conductors. The low temperature form (impurity conduction only) becomes

$$Q = (\ln RT^{\frac{3}{2}} - 5{\cdot}32)k/e + \Delta Q,$$

where R is the Hall constant in cm^3/coulomb and ΔQ is a term to compensate for the dependence of mean-free-path on velocity. In general

$$Q = -\frac{k}{e}\left[38{\cdot}121\,\frac{n_1c - n_2}{n_1c + n_2} - 0{\cdot}659\,\frac{n_2}{n_1c + n_2} - \frac{n_1c}{n_1c + n_2}\ln\left(\frac{n_1}{T^{\frac{3}{2}}}\right) + \frac{n_2}{n_1c + n_2}\ln\left(\frac{n_2}{T^{\frac{3}{2}}}\right)\right]$$

where n_1 and n_2 are the number of electrons and holes per cc determined from conductivity and c is the ratio of electron and hole mobility in the intrinsic range, identical for all germanium samples. The thermoelectric power thus calculated is in agreement with experiment throughout the temperature range.

10

Transition from Classical to Quantum Statistics in Germanium Semiconductors at Low Temperature†‡

ANALYSIS[1] of the experimental results[2] obtained with germanium semiconductors in the temperature range from $-180°C$ to about 600°C has shown that one can account for electrical conductivity and thermoelectric power of these impurity semiconductors by assuming that lattice vibrations and scattering by singly charged impurity centers[3] are responsible for the observed resistivity ρ, where $\rho = \rho_L + \rho_I$.

$$\rho_I = DRT^{3/2},$$

$$\rho_I = \frac{9 \times 10^{11} \pi^{3/2} e^2 m^{1/2}}{2^{7/2} \epsilon^2 (kT)^{3/2}} \ln \left(1 + \frac{36 \epsilon^2 k^2 T^2 d^2}{e^4}\right),$$

where $R \sim 1/n$ is the Hall constant, n the number of conduction electrons per cc, m the electronic mass, ϵ the dielectric constant, $d = 0 \cdot 28 n^{-1/3}$ = one-half the average distance between impurity centers, while the constant D is to be determined from experiments.

In both cases it has been assumed that classical statistics can be applied. This is justified in most cases since the number of electrons, as determined from Hall effect measurements, is small.

If the number of electrons is nearly independent of temperature one may apply the well-known criterion for degeneracy and define a degeneracy temperature

† *Phys. Rev.* **71**, 374, 909 (erratum) (1947), with V. A. Johnson.

‡ This work has been carried out under contract No. 36-039-sc 32020 between the Purdue Research Foundation and the Signal Corps,

$$T_d = \frac{h^2}{8mk}\left(\frac{3n}{\pi}\right)^{2/3} = 4{\cdot}2 \times 10^{-11} n^{2/3}\ ^\circ\mathrm{K}.$$

Since n varies from sample to sample, one finds that degeneracy temperatures vary from a fraction of a degree K to about 150°K in the germanium samples studied at Purdue. Therefore, at low temperatures, the behavior of these semiconductors should vary widely, depending upon the number of electrons and the activation energy.

Measurements of such semiconductors down to about 10°K have been reported recently.[4] The observations show that three kinds of samples exist:

(1) Very pure samples with a resistance increasing so sharply with decreasing temperature that the material becomes almost non-conducting (Estermann's "pure" germanium and silicon samples).

(2) Samples for which the resistivity increases with decreasing temperature and in some cases seems to reach a "saturation" value.

(3) Samples with constant resistivity from liquid air temperature to liquid hydrogen temperature. All of the samples of type (3) have degeneracy temperatures of about 100°K or higher; calculations using classical statistics, such as have been used at medium temperatures, are not justified for such samples at low temperatures.

We have, therefore, carried out calculations assuming Fermi statistics instead of classical statistics and can summarize our results as follows:

(*a*) Lattice scattering.[5]

Above T_d, $\rho_L = DRT^{3/2}$,

Below T_d, $\rho_L = D'RTG(\Theta/T) \rightarrow D'RT^5$ at 25°K.

These expressions, calculated for germanium samples, show a smoothly decreasing resistivity with decreasing temperature and, therefore, contribute little to the observed resistivity at low temperatures.

(*b*) Impurity scattering. By calculating the scattering of electrons by randomly distributed, singly-charged impurity centers, one obtains:

$$\frac{1}{\rho_I} = \sigma_I = \frac{32}{3} \frac{\epsilon^2 m k^3 T^3}{n e^2 h^3} \int_0^\infty \frac{x^3 \exp(x - \mu^*) dx}{[\exp(x - \mu^*) + 1]^2 \ln Y},$$

$$Y = 1 + \frac{4\epsilon^2 k^2 T^2 d^2 x^2}{e^4}, \quad x = \frac{mv^2}{2kT}, \quad \mu^* = \frac{\mu}{kT}$$

where μ is the chemical potential. As T approaches 0, Y can be developed and the integral evaluated, giving finally:[6]

$$\begin{aligned} \rho_I &= (3\pi^2 n)^{1/3} d^2 h/e^2 \text{ e.s.u.} \\ &= 6270\, n^{-1/3} \text{ ohm-cm.} \end{aligned}$$

A detailed theory has to be based on a knowledge of both the Hall constant and the resistivity throughout the temperature range, but comparison of our calculations with Estermann's resistivity values indicates good agreement between theory and experiment (see Table 1).

TABLE 1

Sample	Measured by Estermann	Measured at Purdue	Calculated
26 *Z*	0·0051	0·0044	0·0040
11 *R*	0·0040	0·0034	0·0037
26 *E*	0·0037	0·0033	0·0034
27 *L*	0·0034	0·0029	0·0033

(All of the above values represent constant low temperature resistivities measured in ohm-cm.)

Thus the transition from classical to quantum statistics leads to a constant residual resistance due to impurity scattering in degenerate samples in agreement with experiment.

NOTES

[1] K. Lark-Horovitz and V. A. Johnson, *Phys. Rev.* **69**, 258, 259 (1946).
[2] K. Lark-Horovitz, A. E. Middleton, E. P. Miller, and I. Walerstein,

Phys. Rev. **69,** 258 (1946). K. Lark-Horovitz, A. E. Middleton, E. P. Miller, W. W. Scanlon, and I. Walerstein, *Phys. Rev.* **69,** 259 (1946).

[3] E. Conwell and V. F. Weisskopf, *Phys. Rev.* **69,** 258 (1946).

[4] G. L. Pearson and W. Shockley, *Bull. Am. Phys. Soc.* **21,** 9 (1946). I. Estermann, A. Foner, and J. A. Randall, *Bull. Am. Phys. Soc.* **22,** 31 (1947).

[5] $G(\Theta/T)$ is the Gruneisen function; Mott and Jones, *Properties of Metals and Alloys*, p. 261.

[6] It can be seen that ρ_I is of this form at low temperatures by developing the classical formula for ρ_I and setting $kT = (3n/\pi)^{2/3} h^2/8m$.

11

Theory of Thermoelectric Power in Semiconductors with Application to Germanium†‡

Abstract

The thermoelectric power Q of a semiconductor is found by calculating the Thomson coefficient σ_T from electrical and thermal current density expressions and then integrating the relation $\sigma_T = T\,dQ/dT$. This procedure yields a general expression for Q in terms of the Fermi level, forbidden band width, temperature, ratio of electron to hole mobility, and effective electron and hole masses. In the impurity range the general formula for Q reduces to a simple dependence on the Hall coefficient and temperature if carrier scattering is largely due to the lattice of the semiconductor; the same expression may be used with the addition of a correction term when carrier scattering by impurity ions becomes important at the lower temperatures. When both holes and electrons must be considered as carriers, Q can be evaluated at any temperature from the resistivity and Hall coefficient at that temperature. An expression is also obtained for the thermoelectric power of an intrinsic semiconductor in a form depending on the mobility ratio, forbidden band width at 0°K, and the temperature rate of change of this band width. Hall and resistivity data measured for six polycrystalline germanium samples and two silicon samples have been inserted into the theoretical expressions derived in this paper. The thermoelectric power curves so calculated are found to give generally good agreement with the measured curves.

I. Introduction

Lark-Horovitz, Middleton, Miller, Scanlon, and Walerstein[1,2] have measured the thermoelectric power curves of a number of

† *Phys. Rev.* **92**, 226 (1953), with V. A. Johnson.

‡ This work was assisted first by a National Defense Research Committee contract with the Purdue Research Foundation and later by a Signal Corps Contract, and was reported in part to American Physical Society meetings at New York in January, 1946 and at Durham, North Carolina in March, 1953.

aluminum-doped and antimony-doped polycrystalline germanium samples, with carrier densities ranging from 10^{15} per cm^3 to 7×10^{18} per cm^3. The resistivity and Hall coefficient, as well as thermoelectric power, were measured over a temperature range as wide as 78°K to 925°K for some samples. The calculations described in this paper were carried out in an attempt to explain the behavior of the thermoelectric power of a semiconducting sample on the basis of its Hall curve and resistivity.

Early theoretical work on the behavior of semiconductors contains references to thermoelectric power.[3–5] However, the results of these authors are not given in form suitable for comparison with experiment or for prediction of thermoelectric power behavior from measured Hall and resistivity data.

For use in comparison with experiment, a theoretical thermoelectric power expression must be adaptable for application in the impurity, transition, and intrinsic ranges. In the impurity range of temperatures the numbers of intrinsic electrons and holes due to the thermal excitation of electrons from the filled band to the conduction band are negligible compared to the number of conduction electrons excited from impurity donor levels (*n* type) or the number of holes formed by ionization of acceptor levels (*p* type). Thus, one need consider, in the impurity range, only one sign of carriers. Many samples show "exhaustion" in the impurity range, i.e., all of the donors or acceptors become ionized and the number of carriers per cm^3 remains effectively constant with temperature rise until intrinsic conduction becomes important. Low resistivity samples may show "degeneracy" at low temperatures,[6] hence a general theory of the thermoelectric power of semiconductors must provide for the use of Fermi–Dirac statistics where appropriate.

"Transition range" is a suitable term to apply to those temperatures at which one must consider both the intrinsic carriers and the carriers released from impurity levels. In this range one allows for the presence of carriers of both signs, but one cannot take the electron density to be equal to the hole density. The term "intrinsic range" is reserved for those temperatures at which the intrinsic electrons and holes completely swamp the carriers from impurity levels. Under this condition the electron density equals the hole density.

The general thermoelectric power expression, which is now obtained, can be put into special forms applicable to the various temperature ranges. These results are then used to calculate thermoelectric power curves to be compared with measured values.

II. General Expression for the Thermoelectric Power

The thermoelectric power is found by obtaining the Thomson coefficient from the thermal and electrical current densities and then integrating the appropriate Thomson relation.[7] When both holes and electrons are present, the electrical (j_x) and thermal (w_x) current densities may be written:

$$j_x = \frac{4\pi e}{3}\int_0^\infty v_1^3 l_1\left(\frac{\partial f_1^0}{\partial x} - eE_x\frac{\partial f_1^0}{\partial \epsilon_1}\right)dv_1 - \frac{4\pi e}{3}\int_0^\infty v_2^3 l_2\left(\frac{\partial f_2^0}{\partial x} + eE_x\frac{\partial f_2^0}{\partial \epsilon_2}\right)dv_2, \tag{1}$$

and

$$w_x = -\frac{4\pi m_1}{6}\int_0^\infty v_1^5 l_1\left(\frac{\partial f_1^0}{\partial x} - eE_x\frac{\partial f_1^0}{\partial \epsilon_1}\right)dv_1 - \frac{4\pi m_2}{6}\int_0^\infty v_2^5 l_2\left(\frac{\partial f_2^0}{\partial x} + eE_x\frac{\partial f_2^0}{\partial \epsilon_2}\right)dv_2, \tag{2}$$

where subscript 1 refers to electrons and subscript 2 to holes. Furthermore, no magnetic field is applied, the electric field intensity and temperature gradient possess only X components, l denotes mean free path, f^0 the unperturbed distribution function, v the carrier velocity, ϵ the carrier kinetic energy ($mv^2/2$), and m the effective mass.

The unperturbed distribution functions are

$$f_1^0 = 2m_1^3h^{-3}\left\{1 + \exp\left(\frac{\epsilon_1 - \zeta_1}{kT}\right)\right\}^{-1}, \tag{3a}$$

and

$$f_2^0 = 2m_2^3h^{-3}\left\{1 + \exp\left(\frac{\epsilon_2 - \zeta_2}{kT}\right)\right\}^{-1}, \tag{3b}$$

where ζ_1 and ζ_2 are the "partial" Fermi levels. The quantities ϵ_1 and ζ_1 are zero at the bottom of the conduction band and have positive values in the conduction band, negative in the forbidden band. The zeros of ϵ_2 and ζ_2 are at the top of the filled band with positive values in the filled band, negative in the forbidden band. The condition for thermal equilibrium requires that the partial Fermi levels be related to ζ, the Fermi level of the sample, by

$$\zeta_1 = \zeta \quad \text{and} \quad \zeta_2 = -E_G - \zeta, \tag{4}$$

where E_G is the width of the forbidden band.

The integrals in Eqs. (1) and (2) are simplified by use of the relations

$$\frac{\partial f_1^0}{\partial x} = -\frac{\partial f_1^0}{\partial \epsilon_1}\frac{dT}{dx}\left\{\frac{\epsilon_1}{T} + T\frac{d}{dT}\left(\frac{\zeta}{T}\right)\right\} \tag{5a}$$

and

$$\frac{\partial f_2^0}{dx} = -\frac{\partial f_2^0}{\partial \epsilon_2}\frac{dT}{dx}\left\{\frac{\epsilon_2}{T} - T\frac{d}{dT}\left(\frac{E_G + \zeta}{T}\right)\right\}. \tag{5b}$$

Now Eqs. (1) and (2) may be rewritten as

$$i_x = -\frac{8\pi e}{3m_1^2}\left[\left\{eE_x + T\frac{d}{dT}\left(\frac{\zeta}{T}\right)\frac{dT}{dx}\right\}L_1(1) + \frac{1}{T}\frac{dT}{dx}L_2(1)\right] + \frac{8\pi e}{3m_2^2}\left[\left\{-eE_x - T\frac{d}{dT}\left(\frac{E_G + \zeta}{T}\right)\frac{dT}{dx}\right\}L_1(2) + \frac{1}{T}\frac{dT}{dx}L_2(2)\right] \tag{6}$$

and

$$w_x = \frac{8\pi}{3m_1^2}\left[\left\{eE_x + T\frac{d}{dT}\left(\frac{\zeta}{T}\right)\frac{dT}{dx}\right\}L_2(1) + \frac{1}{T}\frac{dT}{dx}L_3(1)\right] + \frac{8\pi}{3m_2^2}\left[\left\{-eE_x - T\frac{d}{dT}\left(\frac{E_G+\zeta}{T}\right)\frac{dT}{dx}\right\}L_2(2) + \frac{1}{T}\frac{dT}{dx}L_3(2)\right], \tag{7}$$

where

$$L_j(1) = \int^{\infty} \epsilon^j l_1 \frac{\partial f_1^0}{\partial \epsilon}\, d\epsilon \tag{8}$$

and $L_j(2)$ is the corresponding integral containing l_2 and f_2^0.

These current density equations are used to find the Thomson coefficient σ_T from the expression for the rate of heat development per unit volume:

$$dH/dt = E_x j_x - \partial w_x/\partial x = \rho j_x^2 - \sigma_T j_x \frac{dT}{dx} + \frac{d}{dx}\left(\kappa_{el}\frac{dT}{dx}\right), \tag{9}$$

where ρ is the electrical resistivity and κ_{el} the portion of the thermal conductivity due to electron transport. The development of dH/dt yields the result

$$\sigma_T = -\frac{T}{e}\frac{d}{dT}\left(\frac{1}{T}\frac{g_1}{g_2}\right) + \frac{T}{e}\left(\frac{g_3}{g_2}\right), \tag{10}$$

where the functions g_1, g_2, and g_3 are defined by

$$g_1 = L_2(1)/m_1^2 - L_2(2)/m_2^2,$$

$$g_2 = L_1(1)/m_1^2 + L_1(2)/m_2^2,$$

and

$$g_3 = \frac{L_1(1)}{m_1^2}\frac{d}{dT}\left(\frac{\zeta}{T}\right) + \frac{L_1(2)}{m_2^2}\frac{d}{dT}\left(\frac{E_G+\zeta}{T}\right).$$

One of the Thomson relations states that, if Q is the thermoelectric power in a semiconductor–metal circuit with junctions at temperatures T and $T + dT$, the product $T dQ/dT$ equals the difference between the Thomson coefficients of the semiconductor and metal. In this derivation, the Thomson coefficient of the metal is taken as zero, and so

$$\sigma_T = T\, dQ/dT \tag{11}$$

is the equation for determining thermoelectric power. Although the following thermoelectric power values are thus found for the semiconductor relative to a metal of zero Thomson coefficient, the values are approximately applicable relative to any metal because metals have σ_T values very much smaller than those of most semiconducting samples. The sign of Eq. (11) is consistent with the convention that the thermoelectric power is positive if conventional current flows from the semiconductor to the reference metal at the cold junction.

III. The Intrinsic Range

At temperatures high enough that the effects of impurity atoms may be neglected and only intrinsic carriers considered, one can assume that:

(*A*) Classical statistics apply and the distribution functions of Eqs. (3a) and (3b) may be replaced by

$$f_1^0 = 2m_1^3 h^{-3} \exp\{(-\epsilon_1 + \zeta)/kT\}, \tag{12a}$$

$$f_2^0 = 2m_2^3 h^{-3} \exp\{(-\epsilon_2 - E_G - \zeta)/kT\}. \tag{12b}$$

(*B*) The only important scattering of carriers is due to the lattice,[8] and hence the mean free paths l_1 and l_2 are independent of the energies ϵ_1 and ϵ_2.

(*C*) The conduction electron and hole densities are equal, i.e., $n_1 = n_2$. Thence,

$$\zeta = -\tfrac{1}{2}E_G - \tfrac{3}{4}kT\ln(m_1/m_2). \tag{13}$$

Assumptions A and B lead to the following values for the integrals appearing in Eq. (10):

$$\begin{aligned} L_1(1) &= -\,2l_1m_1^3h^{-3}kT\exp(\zeta/kT); \\ L_1(2) &= -\,2l_2m_2^3h^{-3}kT\exp\{-\,(E_G+\zeta)/kT\}; \\ L_2(1) &= -\,4l_1m_1^3h^{-3}(kT)^2\exp(\zeta/kT); \\ L_2(2) &= -\,4l_2m_2^3h^{-3}(kT)^2\exp\{-\,(E_G+\zeta)/kT\}. \end{aligned} \tag{14}$$

Also, when assumptions A and B are valid, the mean free paths can be expressed in terms of the electron and hole mobilities, μ_1 and μ_2, respectively:

$$l_{1,2} = \tfrac{3}{4}\,\mu_{1,2}\,(2\pi m_{1,2}kT)^{\frac{1}{2}}/e. \tag{15}$$

Insertion of expressions (14) and (15) into Eq. (10) and simplification yields a Thomson coefficient expression valid at high temperatures:[9]

$$\sigma_T = -\,\frac{T}{e}\frac{d}{dT}(2k\tanh z) + \frac{T}{e}\left\{-\,\frac{1}{2}\tanh z\,\frac{d}{dT}\left(\frac{E_G}{T}\right) + k\,\frac{dz}{dT}\right\}, \tag{16}$$

where z is defined by

$$z = \frac{\zeta}{kT} + \frac{E_G}{2kT} + \frac{1}{2}\ln\left(\frac{cm_1^{\frac{3}{2}}}{m_2^{\frac{3}{2}}}\right), \tag{17}$$

in which c denotes the mobility ratio μ_1/μ_2.

For an intrinsic semiconductor, one introduces Eq. (13) into Eq. (17) to obtain $z = \frac{1}{2}\ln c$ and thence

$$\sigma_T = -\,\frac{T}{e}\frac{d}{dT}\left\{\frac{2k(c-1)}{c+1}\right\} + \frac{T}{e}\frac{k(c-1)}{(c+1)}\frac{d}{dT}\left(\frac{-\,E_G}{2kT}\right). \tag{18}$$

Now Eq. (11) may be readily integrated to yield for the thermoelectric power

$$Q = -\frac{k(c-1)}{e(c+1)}\left(\frac{E_G}{2kT}+2\right). \tag{19}$$

Both experimental and theoretical considerations[10-14] indicate that E_G varies with temperature in an approximately linear manner:

$$E_G = E_0 + aT. \tag{20}$$

Thence Eq. (19) becomes

$$Q = -\frac{k}{e}\frac{(c-1)}{(c+1)}\left(\frac{E_0}{2kT}+2+\frac{a}{2k}\right). \tag{21}$$

This predicts that a plot of Q as a function of the reciprocal of the temperature, in the intrinsic range, should be a straight line with parameters determined by E_0, a and c. Figure 1 shows a plot of Q vs. $1/T$, at high temperatures, for several of the n-type germanium samples investigated by Middleton and Scanlon. The approach of these curves to a common straight line may be seen; the empirical equation of this line is

$$Q = -\frac{k}{e}\left(\frac{2430°}{T}-0{\cdot}34\right). \tag{22}$$

One can determine a and c for germanium by comparing Eqs. (21) and (22) and inserting an experimentally determined value [1,12,13,15] of E_0. The results are shown in Table 1.

The value of a obtained in this manner is in good agreement with the value obtained for germanium from optical data,[12] but higher than the value found from other data.[11-13] The c values of Table 1 are much larger than values found by methods[16-18] which employ data taken at lower temperatures (usually 300°K) and which give c between 1·5 and 2·1. Such a temperature difference in c values is to be anticipated if the electron mobility follows the expected $T^{-1 \cdot 5}$ law[8] while the hole mobility varies with temperature about as $T^{-2 \cdot 2}$, as has been indicated by recent experiments.[18,19]

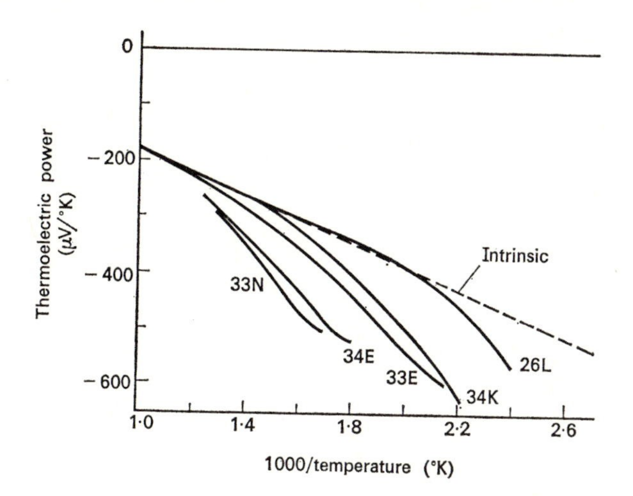

FIG. 1. Thermoelectric power as a function of reciprocal temperature for five n-type, antimony-doped polycrystalline germanium samples. The dashed line is drawn in to approximate the common line approached by all samples as they become intrinsic. The number of conduction electrons per cm^3 at exhaustion is $3 \cdot 3 \times 10^{15}$ for $26L$, $7 \cdot 7 \times 10^{16}$ for $34K$, $1 \cdot 1 \times 10^{17}$ for $33E$, $6 \cdot 2 \times 10^{17}$ for $34E$, and $8 \cdot 2 \times 10^{17}$ for $33N$. The three purer samples become intrinsic within the range of investigation and so empirically determine the curve for intrinsic germanium as $Q = -86 \cdot 3(2430°/T - 0 \cdot 34)\ \mu V/°K$.

TABLE 1. VALUES OF ELECTRON MOBILITY TO HOLE MOBILITY RATIO (c) AND TEMPERATURE VARIATION OF FORBIDDEN BAND WIDTH (a) FOUND FROM THE MEASURED THERMOELECTRIC POWER CURVE OF INTRINSIC GERMANIUM

E_0 (eV)	$c = \mu_1/\mu_2$	a (eV/°K)
0·70	3·98	$-4 \cdot 43 \times 10^{-4}$
0·72	3·80	$-4 \cdot 45 \times 10^{-4}$
0·74	3·62	$-4 \cdot 48 \times 10^{-4}$
0·76	3·47	$-4 \cdot 52 \times 10^{-4}$

IV. THE TRANSITION RANGE

In the transition range, both holes and electrons are present as carriers, but the carriers released by impurities are comparable with those due to intrinsic conduction; thus one cannot take n_1 equal to n_2. However, the temperature is high enough that assumptions A and B and, hence, Eq. (16) are still valid. When Eq. (16) is inserted into Eq. (11) and the integration performed, one obtains

$$Q = -\frac{2k}{e}\tanh z + \frac{k}{e}z - \frac{k}{e}\int \tanh z \frac{d}{dT}\left(\frac{E_G}{2kT}\right)dT$$

$$\approx -\frac{2k}{e}\tanh z + \frac{1}{eT}\left(\zeta + \frac{E_G}{2}\right) - \frac{E_G}{2eT}\tanh z. \qquad (23)$$

The approximation made in evaluating the integral above is that of taking $\tanh z$ as a slowly varying function of T in comparison with $(d/dt)\,(E_G/2kT)$. The error from this approximation vanishes at the low temperature end of the transition range, but increases to $\frac{3}{4}(k/e)\ln(m_2/m_1)$ as the sample becomes intrinsic. The measurements of Benedict and Shockley[20] indicate that m_2/m_1 is probably less than 1·6 for germanium. Hence the maximum difference between the exact and approximate forms of Eq. (23) is probably less than 30 μV/°C.

Equations (12a) and (12b) can be used to convert Eq. (23) into an expression for calculating the thermoelectric power from the electron and hole densities:

$$Q = -\frac{k}{e(n_1c + n_2)}\left[2(n_1c - n_2) - n_1c\ln\left\{\frac{n_1h^3}{2(2\pi m_1kT)^{\frac{3}{2}}}\right\} + n_2\ln\left\{\frac{n_2h^3}{2(2\pi m_2kT)^{\frac{3}{2}}}\right\}\right]. \qquad (24)$$

The comparison of measured thermoelectric power with values calculated from Eq. (24) is discussed in section VI.

V. The Impurity Range

At the lower temperatures the number of intrinsic carriers is negligible compared to the number of carriers released by impurities. When the terms pertaining to hole conduction are dropped from Eq. (10), one obtains

$$\sigma_T(n\text{-type}) = -\frac{T}{e}\frac{d}{dT}\left\{\frac{1}{T}\frac{L_2(1)}{L_1(1)}\right\} + \frac{T}{e}\frac{d}{dT}\left(\frac{\zeta}{T}\right), \tag{25a}$$

and dropping out the electron-conduction terms in Eq. (10) leaves:

$$\sigma_T(p\text{-type}) = \frac{T}{e}\frac{d}{dT}\left\{\frac{1}{T}\frac{L_2(2)}{L_1(2)}\right\} + \frac{T}{e}\frac{d}{dT}\left(\frac{E_G + \zeta}{T}\right). \tag{25b}$$

Equation (11) may be integrated to yield the results,

$$Q(n\text{-type}) = -\frac{1}{eT}\frac{L_2(1)}{L_1(1)} + \frac{\zeta}{eT}, \tag{26a}$$

and

$$Q(p\text{-type}) = \frac{1}{eT}\frac{L_2(2)}{L_1(2)} + \frac{E_G + \zeta}{eT}. \tag{26b}$$

For most germanium, silicon, and tellurium samples, the carriers obey classical statistics above liquid air temperature. In this case the distribution functions are given by Eqs. (12a) and (12b), and the Fermi level may be related to the carrier density by the equation:

$$\zeta = kT \ln \left\{\frac{n_1 h^3}{2(2\pi m_1 kT)^{\frac{3}{2}}}\right\}. \tag{27}$$

If most of the carrier scattering is due to the lattice, the mean-free-path is independent of the carrier kinetic energy, and Eqs. (14) and (27) may be used to obtain

$$L_2(1)/L_1(1) = L_2(2)/L_1(2) = 2kT. \tag{28}$$

Inserting this result into Eqs. (26a) and (26b) produces the expression:

$$Q = \pm \frac{k}{e}\left[2 - \ln\left\{\frac{n_{1,2}h^3}{2(2\pi m_{1,2}kT)^{\frac{3}{2}}}\right\}\right], \tag{29}$$

where the sign of Q is the sign of the carrier. Equation (24), when either n_1 or n_2 is set equal to zero, reduces to Eq. (29).

Under the conditions assumed in deriving Eq. (29), the carrier density is related to the Hall coefficient of the sample, R, by the relation:[21]

$$n = 3\pi/(8e|R|), \tag{30}$$

where R is measured in cm^3/coulomb, e in coulomb, and n per cm^3. When Eq. (30) is substituted into Eq. (29) along with the values of the various quantities, including the free electron mass for $m_{1,2}$, a simple thermoelectric power expression is obtained:

$$Q = \pm (k/e)[\ln(|R|T^{\frac{3}{2}}) - 5{\cdot}32]. \tag{31}$$

When this expression is valid, the thermoelectric power as a function of temperature may be found from a measured Hall curve which gives R as a function of T; and, conversely, measurement of the thermoelectric power curve gives an approximate determination of the Hall curve and carrier density curve of the same sample.

The assumption of predominant lattice scattering is best satisfied for relatively high purity samples at the higher temperature end of the impurity range. It has been found, especially in germanium, that, with decreasing temperature and increasing impurity content, a substantial portion of the carrier scattering is caused by the randomly distributed impurity ions.[22,23] An approximate correction for the effect of impurity scattering upon the thermoelectric power is based upon the result of Conwell and Weisskopf[22] that the mean free path due to impurity scattering is about proportional to the square of the kinetic energy, whereas Sommerfeld and Bethe[8] have found the mean free path due to lattice scattering (l_L) to be independent of energy. When both kinds of scattering are present, the mean-free-path l is given by

$$1/l = 1/l_L + 1/l_I. \tag{32}$$

If $l_I = a\epsilon^2$, the energy dependence of l is given by

$$l = \frac{al_L\epsilon^2}{l_L + a\epsilon^2}. \tag{33}$$

This expression enters into the determination of the Hall coefficient so that Eq. (30) should be replaced by

$$n = r/e|R|, \tag{34}$$

where r depends[24,25] on the ratio of a/l_L (or ρ_L/ρ_I) in the manner given in Table 2.[26]

TABLE 2. DEPENDENCE OF THE HALL COEFFICIENT FACTOR r, DEFINED AS $ne|R|$, AND THERMOELECTRIC POWER UPON RELATIVE PROPORTIONS OF LATTICE AND IMPURITY SCATTERING

$\frac{\rho_I}{\rho_I + \rho_L}$	$\frac{\rho_I}{\rho_L}$	r	$q = \frac{L_2}{kTL_1}$
0·0	0	1·1781	2·000
0·1	1/9	1·0289	2·379
0·2	1/4	1·0411	2·558
0·3	3/7	1·0650	2·698
0·4	2/3	1·0962	2·809
0·5	1	1·1348	2·947
0·6	3/2	1·1833	3·065
0·7	7/3	1·2475	3·199
0·8	4	1·3399	3·380
0·9	9	1·4963	3·639
1·0	∞	1·9328	4·000

The quantity $q = L_2(1)/\{kTL_1(1)\} = L_2(2)/\{kTL_1(2)\}$ also depends upon the relative amounts of lattice and impurity scattering in a manner found by inserting Eq. (33) into Eq. (8); the results of this computation are also given in Table 2. When the corrections due to impurity scattering are considered and allowance is made for the difference between free electron mass m_0 effective mass, Eq. (31) is replaced by

$$Q = \pm (k/e)[\ln(|R|T^{\frac{3}{2}}) - \ln r - 7 \cdot 16 + q + \tfrac{3}{2} \ln(m_{1,2}/m_0)]. \tag{35}$$

Since modifications due to using Fermi–Dirac statistics in place of Boltzmann statistics are usually not important above liquid air temperature, thermoelectric power in degenerate semiconductors is not discussed here but will be presented in a later paper.

VI. Comparison with Experiment

The theoretical expressions developed in the preceding sections have been compared with the thermoelectric power curves measured by Middleton and Scanlon.[2]

The thermoelectric power curves in the impurity range are calculated by applying Eq. (31), or Eq. (35) where required, to the measured Hall curves of the samples. Equation (35) is used for quite impure samples at relatively low temperatures; m_1 and m_2 are taken equal to m_0, but proper values of r and q are inserted on the basis of analysis[27] of the resistivity curves of the samples.

The theoretical curves in the transition range are computed by putting into Eq. (24) values obtained from measured Hall and resistivity curves.[2] The carrier densities n_1 and n_2 are found, for a given temperature, from the values of R and σ (electrical conductivity) at that temperature by using the equations:

$$R = -\frac{3\pi}{8e} \frac{n_1 c^2 - n_2}{(n_1 c + n_2)^2}, \tag{36}$$

$$\sigma = n_1 e \mu_1 + n_2 e \mu_2 = 1/\rho, \tag{37}$$

and

$$n_2 - n_1 = \pm N, \tag{38}$$

where N is the number of carriers per cm^3 in the exhaustion range (the $+$ sign applies to p-type samples, the $-$ sign to n-type).

Algebraic elimination of n_1 and n_2 yields expressions for the mobility ratio c:

$$(n\text{-type})\ 1 - \frac{1}{c} = -\left\{\frac{8R}{3\pi e}\left(\frac{\sigma}{\mu_1}\right)^2 + N\right\}\left(\frac{\sigma}{\mu_1 e} - N\right)^{-1}, \tag{39a}$$

and

$$(p\text{-type})\ c - 1 = \left\{N - \frac{8R}{3\pi e}\left(\frac{\sigma}{\mu_2}\right)^2\right\}\left(\frac{\sigma}{\mu_2 e} - N\right)^{-1}. \tag{39b}$$

Only lattice scattering is important in the transition range, and so the electron mobility as a function of temperature[8] is given by

$$\mu_1 = BT^{-\frac{3}{2}}, \tag{40}$$

where B is about $1\cdot 8 \times 10^7\ {}^\circ\mathrm{K}^{\frac{3}{2}}\ \mathrm{cm^2/volt\text{-}sec}$ for electrons in single crystal germanium and less for the polycrystalline samples investigated by Middleton and Scanlon. For these polycrystalline samples mobility was extrapolated, following Eq. (40), from values at the high temperature end of the impurity range. The extrapolated mobility and measured R and σ at a chosen T are put into Eq. (39a) or (39b) to determine c at that temperature for the chosen sample. This process yielded c values for germanium, at temperatures in the 600°–900°K range, that averaged to about 3·0. No systematic variation of c with temperature or impurity content was observed. When c has been found, n_1 and n_2 may be evaluated from Eqs. (37) and (38), thus completing the data required for evaluation of thermoelectric power from Eq. (24).

Figure 2 shows a comparison between the measured and calculated thermoelectric power curve for three n-type, antimony-doped polycrystalline germanium samples and three p-type, aluminum-doped polycrystalline germanium samples. It is apparent that consistently good agreement exists in the transition range, where there is little scatter of the experimental points, and also quite good agreement in the impurity range in view of the rather wide scatter of the experimental points at these lower temperatures.

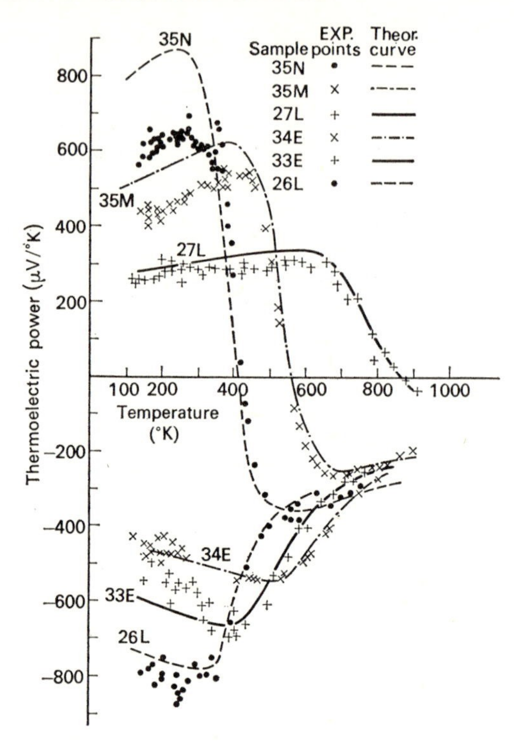

FIG. 2. Comparison of calculated and measured thermoelectric power curves for polycrystalline germanium samples. The smooth curves are calculated, and experimental points are indicated by dots and crosses. The number of holes per cm³ at exhaustion is $5 \cdot 7 \times 10^{15}$ for 35*N*, $1 \cdot 7 \times 10^{17}$ for 35*M*, and $7 \cdot 2 \times 10^{18}$ for 27*L*; the number of electrons per cm³ at exhaustion is $3 \cdot 3 \times 10^{15}$ for 26*L*, $1 \cdot 1 \times 10^{17}$ for 33*E*, and $6 \cdot 2 \times 10^{17}$ for 34*E*.

Figure 3 shows a similar comparison between theory and experiment for two polycrystalline silicon samples, both *p* type, one (26*G*) boron-doped and the other (112) aluminum-doped. While there is a fair degree of agreement between theory and experiment, it is not as good as for the germanium samples, perhaps because all measurements on silicon were more difficult

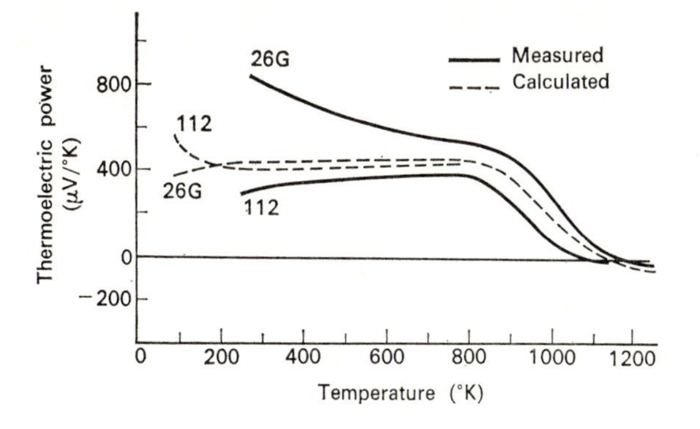

FIG. 3. Comparison of calculated and measured thermoelectric power curves for two *p*-type silicon samples, aluminum-doped 112 and boron-doped 26*G*.

than on germanium because pressure contacts were used instead of soldered ones.

The authors wish to thank A. E. Middleton, W. W. Scanlon, E. P. Miller, and I. Walterstein for their courtesy and cooperation in making available all of their experimental results continuously throughout the progress of this investigation.

NOTES

[1] Lark-Horovitz, Middleton, Miller, Scanlon, and Walerstein, *Phys. Rev.* **69,** 259 (1946).

[2] A. E. Middleton and W. W. Scanlon, *Phys. Rev.* **92,** 219 (1953).

[3] M. Bronstein, *Physik. Z. Sowjetunion* **2,** 28 (1932).

[4] R. H. Fowler, *Proc. Roy. Soc.* (*London*) **A140,** 505 (1933).

[5] A. H. Wilson, *Theory of Metals* (Cambridge University Press, Cambridge, England, 1936), p. 181.

[6] V. A. Johnson and K. Lark-Horovitz, *Phys. Rev.* **71,** 374, 909 (1947).

[7] See, e.g., F. Seitz, *The Modern Theory of Solids* (McGraw-Hill Book Company, Inc., New York, 1940), p. 174.

[8] A. Sommerfeld and H. Bethe, *Handbuch der Physik* (Verlag Julius Springer, Berlin, Germany, 1933), Vol. 24, No. 2, pp. 509–521, 558–560.

[9] A similar expression is given in a discussion of the thermoelectric power of tellurium samples by T. Fukuroi and S. Tanuma, *Science Repts. Research Insts., Tôhoku Univ.* **4,** 353 (1952).

[10] J. Bardeen, *Phys. Rev.* **75,** 1777 (1949).
[11] W. Shockley and J. Bardeen, *Phys. Rev.* **77,** 407 (1950).
[12] H. Y. Fan, *Phys. Rev.* **78,** 808 (1950).
[13] V. A. Johnson and H. Y. Fan, *Phys. Rev.* **79,** 899 (1950).
[14] T. S. Moss, *Phys. Rev.* **79,** 1011 (1950).
[15] J. R. Haynes and H. B. Briggs, *Phys. Rev.* **86,** 647 (1952).
[16] G. L. Pearson, *Phys. Rev.* **76,** 179 (1949).
[17] L. P. Hunter, *Phys. Rev.* **91,** 207 (1953).
[18] M. B. Prince, *Phys. Rev.* **91,** 208 (1953).
[19] W. C. Dunlap, *Phys. Rev.* **79,** 286 (1950).
[20] T. S. Benedict and W. Shockley, *Phys. Rev.* **91,** 207 (1953).
[21] R. Gans, *Ann. Physik* **20,** 293 (1906).
[22] E. Conwell and V. F. Weisskopf, *Phys. Rev.* **69,** 258 (1946); **77,** 388 (1950).
[23] K. Lark-Horovitz and V. A. Johnson, *Phys. Rev.* **69,** 258 (1946).
[24] H. Jones, *Phys. Rev.* **81,** 149 (1951).
[25] V. A. Johnson and K. Lark-Horovitz, *Phys. Rev.* **82,** 977 (1951).
[26] A later paper will describe corrections required in resistivity, thermoelectric power, and Hall effect by the deviation of l_l from a simple ϵ^2 energy dependence.
[27] K. Lark-Horovitz, *National Defense Research Committee Report* 14–585, pp. 36, Nov. 1945 (unpublished); K. Lark-Horovitz and V. A. Johnson, *Phys. Rev.* **69,** 258 (1946); K. Lark-Horovitz, *Elec. Eng.* **68,** 1047 (1949); H. C. Torrey and C. A. Whitmer, *Crystal Rectifiers* (McGraw-Hill Book Company, Inc., New York, 1948), pp. 58–61 and Figs. 3–7.

12

Impurity Band Conduction in Germanium and Silicon†

The Electrical Properties of Germanium Semiconductors at Low Temperatures‡

ABSTRACT

Low-temperature anomalies in the Hall effect—a steep maximum—and a change of the slope of the log resistivity versus $1/T$ curve in germanium semiconductors, which were first observed by Hung and Gliessman, have been re-investigated using single crystals of N- and P-type germanium of various carrier concentrations. By using cross-shaped samples, the influence of contacts was investigated. To estimate the effect of surface layers, various surface treatments have been tried and the ratio of surface area to bulk volume has been changed. It was found that the effects cannot be explained by surface conduction. To exclude electrical field effects at low temperatures (to about $1 \cdot 50°$K), fields as low as 3 mV/cm, well below breakdown field, have been used. The Hall effect was measured with varying magnetic fields ranging from 200 to 4300 gauss. In the same temperature region where Hall effect and resistivity become anomalous, the magneto-resistive ratio also shows drastic changes. In the temperature range before the Hall maximum is reached the magneto-resistive ratio changes only slowly with temperature, but starts to decrease sharply at the temperature where the Hall effect approaches the maximum and where the resistivity curve changes slope. The discussion of the results indicates that the observations can be described by a model which assumes conduction in two bands, the regular conduction band (valence band in the case of P-type material) and a band with a very small mobility. The sharp decrease of this mobility with decreasing impurity content suggests that conduction in an impurity band is a plausible explanation of these phenomena.

Some time ago Hung and Gliessman[1] observed that the electrical behavior of germanium semiconductors shows anomalies in the

† *Physica* **20,** 834 (1954), with H. Fritzsche, and *J. Phys. Chem. Solids* **8,** 259 (1959), with T. A. Longo and R. K. Ray.

‡ Supported by a U.S. Signal Corps contract.

low temperature range. They investigated chemically doped *N*-type material and three *P*-type samples produced by nuclear transformations with slow neutrons in the nuclear reactor; these *P*-type samples contain gallium and arsenic atoms in the ratio 3 to 1.

At first resistivity and Hall coefficient increase exponentially with decreasing temperature as expected from the simple theory of elementary semiconductors, but at a certain temperature the Hall coefficient reaches a maximum and then decreases sharply by orders of magnitude. In the same temperature range the resistivity seems to reach a saturation value.[2]

One can, therefore, distinguish in the $\lg \varrho$ vs. $1/T$ curve a larger slope in the temperature range before the Hall maximum is reached and a much smaller slope at still lower temperatures.

A striking observation of Hung confirmed by later experiments of Finlayson† is the fact that the maximum of the Hall coefficient is shifted to higher temperatures and decreases in magnitude with increasing number of impurity atoms.

For the explanation of these results Hung[3] postulated conduction in an impurity band in addition to conduction in the ordinary conduction band. By assuming that the number of carriers in the impurity band is equal to the number of carriers in the conduction band observed at higher temperatures in the "exhaustion range" (flat Hall curve), and that the mobility in the impurity band is extremely small compared to the ordinary mobility in the conduction band, he obtained reasonable agreement with the experiments.

The conduction properties of the impurity band have been discussed by Schottky,[4] Busch,[5] James and Ginzbarg,[6] Gisolf,[7] Erginsoy,[8] and recently by Baltensperger.[9]

In any theoretical discussion the explanation of the formation of a conduction band with impurity concentrations as low as 10^{15}/cc is rather difficult.

The experiments have, therefore, been repeated, extending the range of observation to about 1·5°K by use of more sensitive

† Private communication.

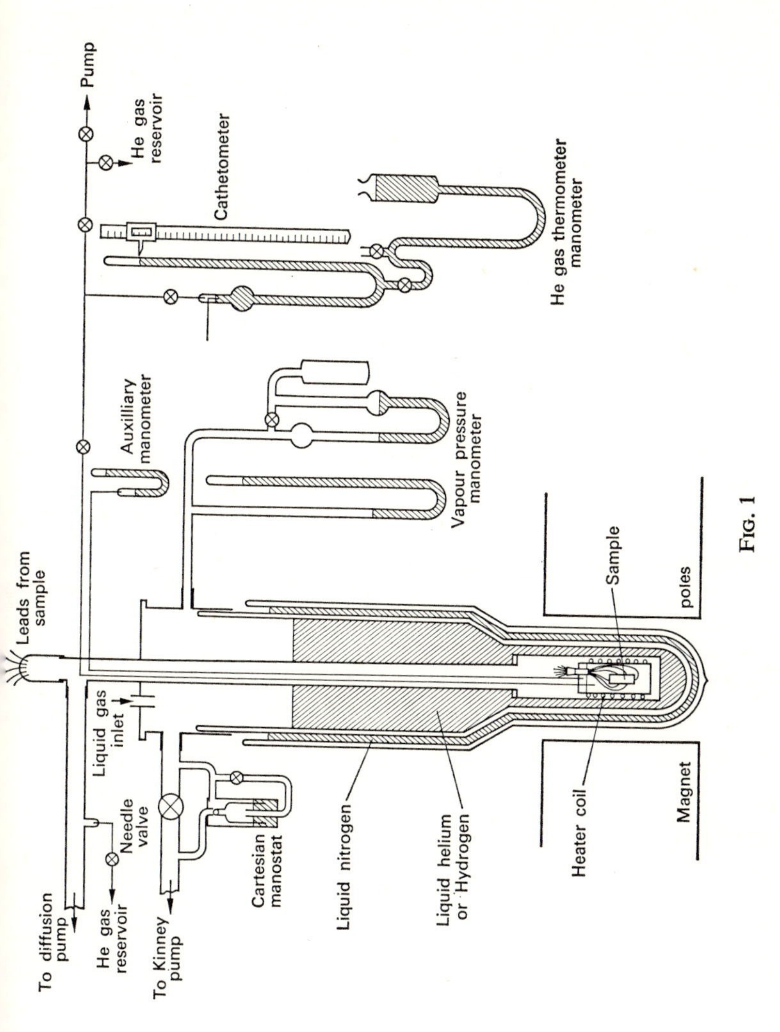
Pump
He gas reservoir
Cathetometer
He gas thermometer manometer
Auxilliary manometer
Vapour pressure manometer
Leads from sample
Liquid gas inlet
Needle valve
To diffusion pump
He gas reservoir
To Kinney pump
Cartesian manostat
Liquid nitrogen
Liquid helium or Hydrogen
Heater coil
Magnet
poles
Sample

FIG. 1

detecting devices. Single crystals were produced with appropriate amounts of indium, gallium, and antimony added to pure germanium. The indium-doped crystals were cut from different ingots. All gallium-doped crystals shown in Fig. 5 and Fig. 6 were cut from the same melt, cutting the plate normal to the axis of the ingot and thus to the concentration gradient.

The experiment arrangement is outlined in Fig. 1. The accuracy of calibration of the helium gas thermometer and, therefore, of the temperature measurement was improved by a factor of 10

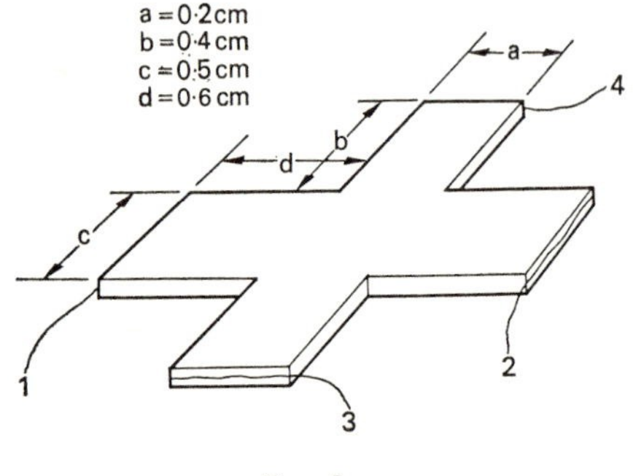

FIG. 2

over previous arrangements. A vibrating reed electrometer allows measurements of current of the order 10^{-16} A. To check the effect of the soldered contacts to the leads a special cross-shaped sample was prepared (Fig. 2), and leads were connected to the legs of the cross. After these measurements the cross bars were cut off and ordinary leads soldered on; the results of both measurements agreed in the whole temperature range. To study the effect of surface treatment, ground, etched and electro-polished samples were prepared with various ratios of volume to surface area and with different surface recombination rates (Fig. 3). Resistivity and Hall effect measurements show that in etched samples, resistivity and Hall coefficient are independent of sample thickness. The ground samples show a smaller Hall coefficient and resistivity

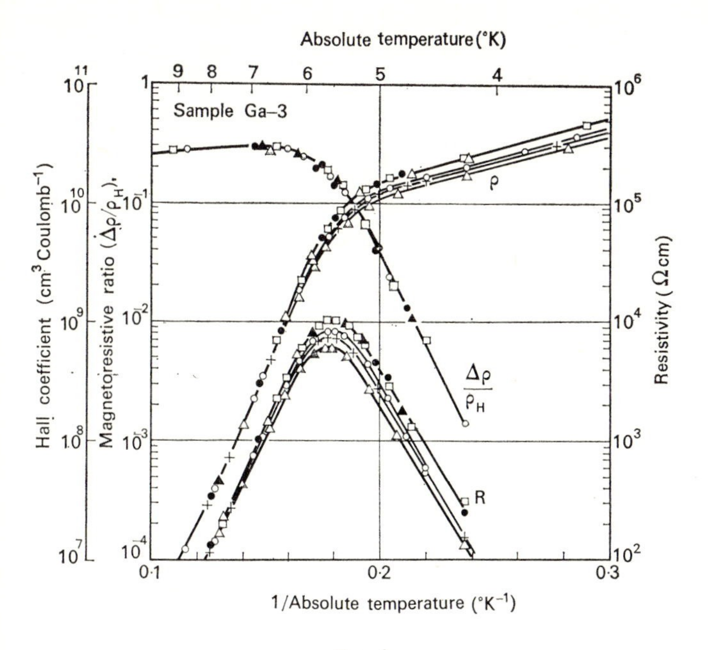

Fig. 3

than the etched samples, and the measured values depend on sample thickness but this effect decreases with increasing temperature because of the decreasing bulk resistivity. The effect of surface treatment, however, is so small that it hardly contributes to the main anomaly which is observed.

The low magnetic field approximation may be used, because measurements with varying magnetic fields show that the shape of the Hall curve is not significantly altered by the magnetic field at which the experiments are made.

Throughout the temperature range investigated, the condition that the Hall electric field, E_y, be small compared to the applied electric field, E_x, was fulfilled.

It has been suggested that the application of an electric field may produce excitation of an avalanche of carriers.[10] This increase would then cause a sharp drop in Hall coefficient and also would produce a change in the slope of the resistivity curve. In all the samples investigated, as listed in Table 1, the behavior throughout the whole temperature range is ohmic (Fig. 4). No field dependence of resistivity was observed between fields of 10^{-3} V/cm and 0·5 V/cm. The decrease of resistivity found to start in fields above

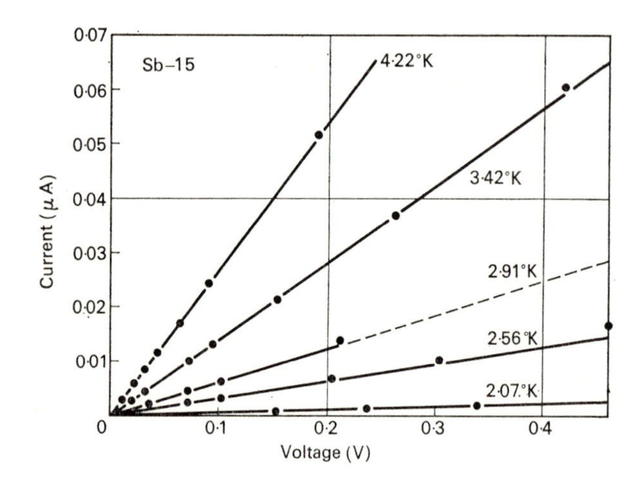

FIG. 4

0·5 V/cm was due to a temperature rise in the sample caused by an increase of sample current. In the same range of applied electric Field, the Hall coefficient also remains unchanged.

The mobilities (see Table 1) in an impurity band are expected to be extremely small, if the impurity concentration is $10^{16}/cm^3$ or smaller. Therefore it is of interest to investigate the magneto-resistive effect, which is in theory directly proportional to the square of the mobility. Measurements with a field of 3500 gauss show a sharp decrease in magneto-resistive ratio at the temperature of the Hall maximum (Fig. 3 and Fig. 7).

GALLIUM OR ANTIMONY. THE HALL COEFFICIENT AT ROOM TEMPERATURE LISTED IN THE 2ND COLUMN IS EQUAL TO R_{ex}, THE VALUE IN THE EXHAUSTION RANGE. IN THE 4TH AND 7TH COLUMN THE HALL MOBILITIES R/ϱ AT THE RESPECTIVE TEMPERATURES ARE LISTED. THE LAST COLUMN SHOWS $R_{ex}/\varrho(T)$, THE MOBILITIES IN THE IMPURITY BAND, AT 2·5°K. THESE VALUES CANNOT BE COMPARED WITH THOSE IN BRACKETS WHICH WERE MEASURED AT 4·2°K

Sample Code Number	$T = 297°K$			$T = 77°K$			$\frac{R_{ex}}{\rho(2\cdot5°K)}$
	R	ϱ	R/ϱ	R	ϱ	R/ϱ	
	cm^3 $coulomb^{-1}$	ohm cm	cm^2 $volt^{-1}$ sec^{-1}	cm^3 $coulomb^{-1}$	ohm cm	cm^2 $volt^{-1}$ sec^{-1}	cm^2 $volt^{-1}$ sec^{-1}
In-1	+ 15300	4·56	3350	+ 10700	0·49	21800	$(1\cdot27 \times 10^{-3})$
In-2	+ 4800	1·67	2870	+ 3250	0·30	10800	(0·16)
In-3	+ 201	0·191	1040	+ 230	0·082	2800	(2·04)
In-4	+ 160	0·0945	1690	+ 190	0·049	3880	0·941
In-5	+ 79·2	0·0785	1000	+ 130	0·054	2400	10·0
Ga-1	+ 3310	1·055	3140	+ 2600	0·122	21300	$(3\cdot70 \times 10^{-3})$
Ga-2	+ 2410	0·809	2980	+ 1900	0·115	16500	$(6\cdot02 \times 10^{-3})$
Ga-3	+ 1800	0·620	2900	+ 1400	0·10	14000	$7\cdot5 \times 10^{-4}$
Ga-4	+ 1325	0·495	2680	+ 1250	0·096	13000	$1\cdot35 \times 10^{-3}$
Ga-5	+ 950	0·375	2540	+ 830	0·074	11200	$1\cdot234 \times 10^{-3}$
Ga-6	+ 410	0·190	2160	+ 435	0·057	7640	$6\cdot03 \times 10^{-3}$
Ga-4-1	+ 380	0·174	2180	+ 420	0·056	7550	$3\cdot55 \times 10^{-3}$
Ga-4-2	+ 240	0·116	2070	+ 280	0·050	5600	$2\cdot0 \times 10^{-2}$
Ga-4-3	+ 176	0·096	1830	+ 224	0·043	5200	$7\cdot0 \times 10^{-2}$
Ga-4-4	+ 120	0·075	1600	+ 160	0·0425	3760	6·49
Sb-15	− 1000	0·285	3510	− 1130	0·074	15270	$5\cdot0 \times 10^{-4}$
Sb-9-21	− 624	0·196	3170	− 760	0·066	11500	$1\cdot6 \times 10^{-3}$

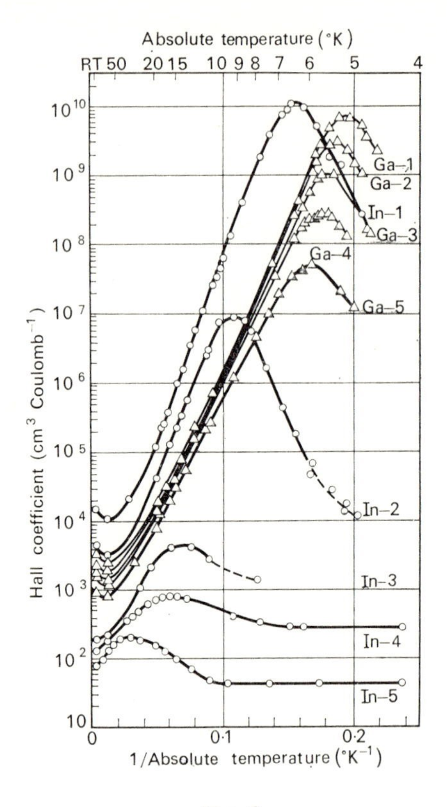

FIG. 5

Upon comparing the results obtained for the resistivity and the Hall coefficient of various samples (Fig. 5 and Fig. 6), it is noticeable that the low temperature behavior observed by Hung has been essentially confirmed.† The family of curves representing Hall effect and resistivity measurements in the gallium-doped samples shows the shift of the onset of the anomaly towards higher temperatures with increasing impurity concentration.

† However not only the resistivity vs. T curve but also the Hall curve shows a definite slope at the lowest temperatures, as recent measurements show.

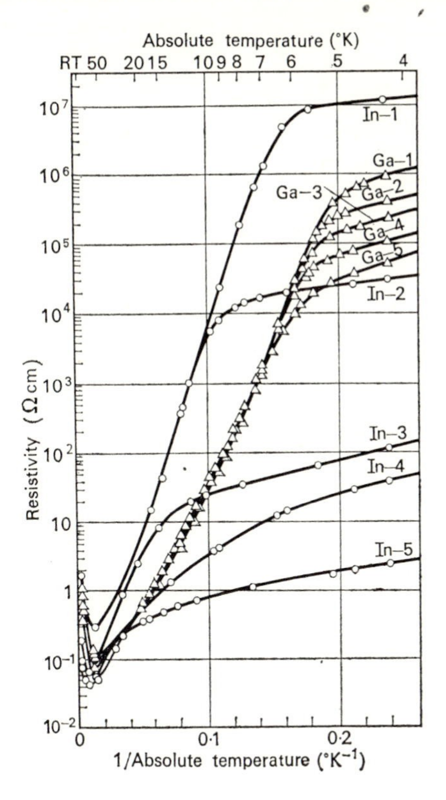

FIG. 6

The more irregular behavior of the indium alloys can be understood, since each sample comes from a different ingot. One also should consider the difference in boiling point between indium and gallium and the difference in the distribution coefficients† of the two metals between liquid and solid germanium. The low boiling point of indium, as compared to the boiling point of gallium, and the low distribution coefficient of indium ($C_S/C_L =$

† The ratio of the concentration of atoms in the solid C_S to concentration of atoms in the liquid C_L.

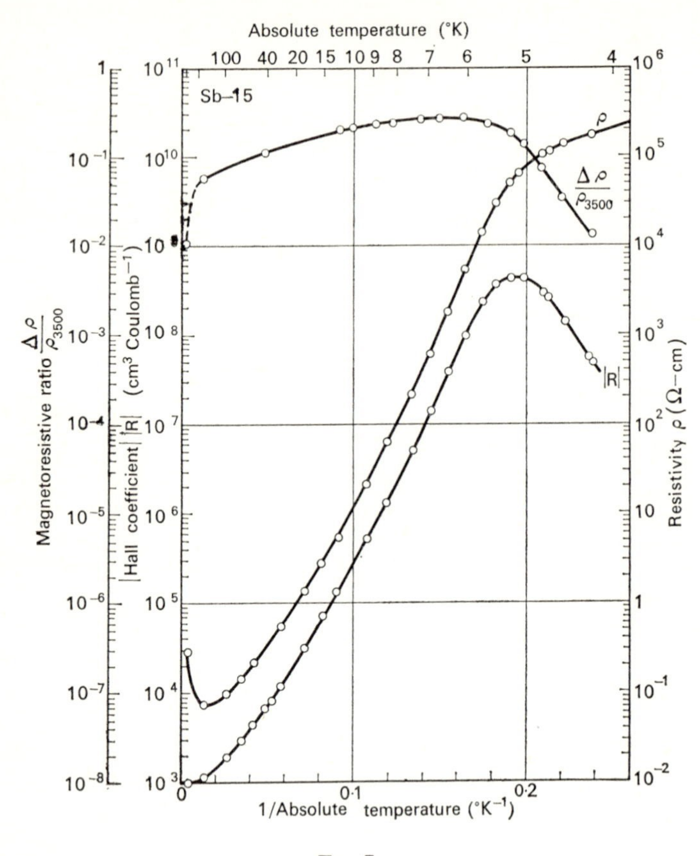

FIG. 7

10^{-3}), as compared to gallium ($C_S/C_L = 0 \cdot 1$),[11] necessitates much larger additions of indium metal than of gallium metal to obtain the same carrier concentration. This also makes it likely that more accidental impurities of a donor-type (N_D) are added to the indium-doped samples than to the gallium-doped samples. Since the Hall effect measures only the density of charge carriers, which in *P*-type samples is $N_A - N_D$, the difference between acceptor,

N_A, and donor densities, N_D, it is not surprising that the Hall maximum of an indium sample with a higher Hall coefficient than the highest one observed in a gallium-doped sample is shifted to higher temperatures by the relatively large impurity content.† The last column of Table 1 shows the values of R_{ex}/ϱ measured at 4·2°K and 2·5°K, as a measure of the mobility in the impurity band. In the gallium-doped samples the Hall constant at exhaustion R_{ex} changes from sample Ga-3 to sample Ga-4-4 by a factor of about 15, the corresponding R_{ex}/ϱ values at room temperature decrease by a factor of approximately 1·7. However the "impurity mobility" (indicated by R_{ex}/ϱ at 2·5°K) changes by a factor of about 10^4. This drastic change is in our opinion one of the strongest arguments in favor of an impurity band conduction.

In addition to these P-type samples, antimony-doped N-type samples have been studied. The measurements made on these samples also agree with Hung's observations (Fig. 7).

Discussion

The experimental arrangement (Fig. 1) excludes the possibility of explaining the effects as due to a heat or light leak from the surroundings. In the very low electric fields applied, it is difficult to picture a breakdown, particularly in view of the ohmic behavior of the sample.

The present experiments and the experiments of Hung and Finlayson indicate that the sign of the carriers is the same throughout the whole temperature range in which observations were made. As James and Ginzbarg[6] have pointed out the number of states in the impurity band is less than twice the number of impurity atoms due to the random distribution of impurities.

Therefore, an impurity band would be more than half-filled by

† We are indebted to V. A. Johnson for an analysis of the log resistivity vs. $1/T$ curves of In-3 and Ga-4-1. The results show that in the indium sample with an effective number of $3 \cdot 7 \times 10^{16}$ holes/cm^3 there are $3 \cdot 8 \times 10^{16}$/cm^3 additional compensating impurities, whereas in the gallium samples with an effective number of holes of 2×10^{16}/cm^3 there are only $2 \cdot 7 \times 10^{15}$/cm^3 additional compensating impurities.

a number of electrons equal to the number of impurity atoms. On this basis one might expect a reversal in the sign of the Hall effect at very low temperatures. However, one has to remember that the theory of conduction and Hall effect in an impurity band, due to impurities distributed at random, has not been worked out and conclusions drawn from expectation for a regular arrangement are somewhat dubious.

We have looked for a reversal in the sign of the Hall effect particularly in measurements of samples which could be carried out at very low temperatures. We have not found any case of a reversal of Hall effect so far. Samples which were *P*-type or *N*-type at room temperature were found to be *P*-type or *N*-type also at low temperature. However if one can assume that there are always impurity atoms of opposite sign present, this is understandable.

One can, therefore, write with Hung:

$$R = (r_c\, en_c\, \mu_c^2 + r_i\, en_i\, \mu_i^2)/(en_c\, \mu_c + en_i\, \mu_i)^2 \tag{1}$$

$$\sigma = (en_c\, \mu_c + en_i\, \mu_i) \tag{2}$$

$$R_c = r_c/en_c \quad R_i = r_i/en_i \tag{3}$$

R_c is the Hall constant in the conduction band, R_i is the Hall constant in the impurity band; r_c and r_i are proportionality constants, dependent on the statistical behavior of the carriers and mean free path dependence on the energy. In a similar way σ_c and σ_i are conductivities in the conduction and impurity band, and ϱ_c and ϱ_i the corresponding resistivities.

Remembering that $n_c + n_i = N = N_A - N_D$, it follows from (1) that the Hall maximum occurs when $n_c\mu_c = n_i\mu_i$. Considering what happens on the low-temperature side of the Hall maximum one can set $n_i = N$, the carrier concentration in the exhaustion range. The mobilities in the impurity band are small compared to the mobilities in the conduction band, even if they do change slowly, therefore one can neglect $n_i\mu_i^2 \ll n_c\mu_c^2$, and one obtains from (1), (2) and (3):

$$R/R_c = \varrho^2/\varrho_c^2 \tag{4}$$

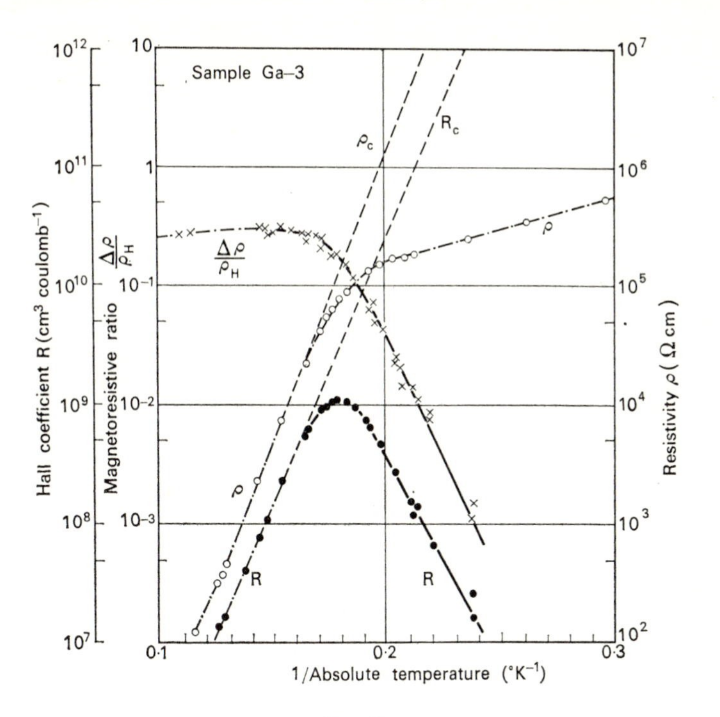

FIG. 8

Since the magneto-resistive ratio is negligible after the Hall maximum is reached one finds†

$$[\varrho(H) - \varrho(0)]/\varrho(H) = \Delta\varrho/\varrho = B_c H^2 \, \varrho/\varrho_c \qquad (5)$$

ϱ_c is calculated from the n_c extrapolated from the Hall curve and μ_c extrapolated from the mobility curve. R and $\Delta\varrho/\varrho$ calculated from ϱ in this way show that this description is self-consistent (Fig. 8).

† $\sigma(H) = \sigma_c(H) + \sigma_i(H)$
$\sigma_c(H) = \sigma_c(0)\,(1 - B_c H^2)$
$\sigma_i(H) = \sigma_i(0)\,(1 - B_i H^2)$
one can neglect $B_i H^2$ as being small compared with unity as found experimentally. Thus one obtains by using resistivities instead of conductivities the formula given in the text.

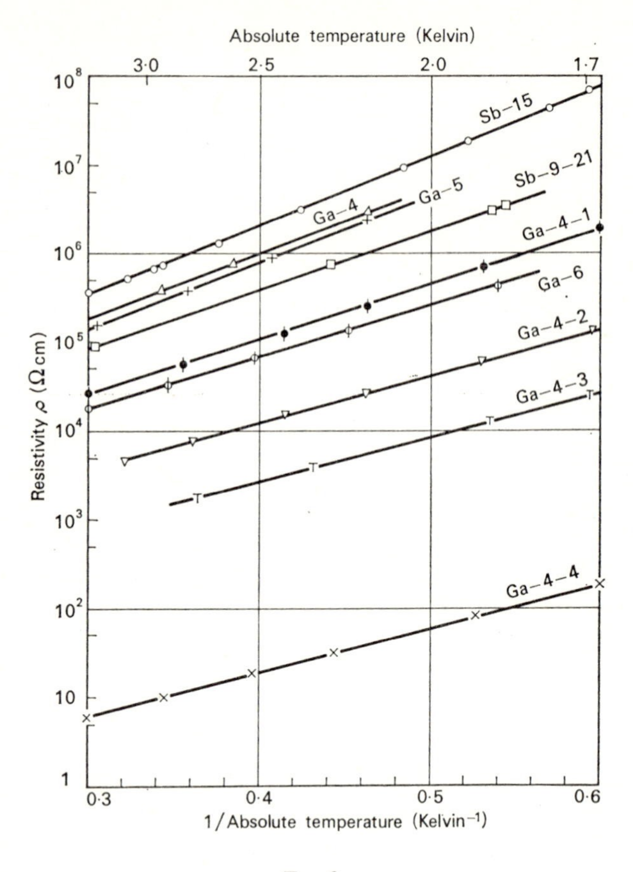

FIG. 9

To check the possible influence of crystal orientation in the magnetic field three plates have been cut out of the same *N*-type ingot to allow different directions of current flow through the crystal [100], [110] with the magnetic field in the direction [001] and [011]. The magneto resistance did not show any change between current *i* in the direction [100] or [110] and *H* [001] as one would expect;[12] there was a difference observed between *i* [110] *H* [001] and *i* [110] *H* [011]. However these differences in orientation do not affect the position of the Hall maximum or the shape of the Hall curve as a function of temperature.

The existence of a small but finite slope of the log ϱ vs. $1/T$ curve in the lowest temperature range has now been established (Fig. 9). This slope may be due to the distribution of carriers over the available states in the impurity band which changes with temperature, and thus the effective mass and mobility may be temperature dependent.

We wish to thank H. Y. Fan and H. M. James for some interesting discussions, we are indebted to Miss L. M. Roth for the preparation of samples and to P. H. Keesom for making the facilities of the cryogenic laboratory available.

REFERENCES

1. C. S. HUNG and J. R. GLIESSMAN, *Phys. Rev.* **79,** 726 (1950).
2. A. N. GERRITSEN, *Physica* **15,** 427 (1949).
3. C. S. HUNG, *Phys. Rev.* **79,** 727 (1950).
4. W. SCHOTTKY, *Z. Elektrochemie* **45,** 33 (1939).
5. G. BUSCH and H. LABHART, *Helv. Phys.* **19,** 463 (1946).
6. H. M. JAMES and A. S. GINZBARG, *J. Phys. Chem.* **57,** 840 (1953).
7. J. H. GISOLF, *Ann. Physik* **1,** 3 (1947).
8. C. ERGINSOY, *Phys. Rev.* **80,** 1104 (1950), also *Phys. Rev.* **88,** 893 (1952).
9. X. BALTENSPERGER, *Phil. Mag.* **44,** 1355 (1953).
10. G. DRESSELHAUS, A. F. KIP and C. KITTEL, *Phys. Rev.* **92,** 827 (1953).
11. R. N. HALL, *J. Phys. Chem.* **57,** 836 (1953).
12. G. L. PEARSON and H. SUHL, *Phys. Rev.* **83,** 768 (1951).

Low Temperature Impurity Conduction in Silicon

1. INTRODUCTION

Impurity conduction has been observed in silicon samples containing B, Al, As or Sb. The measurements were made using standard techniques for measuring Hall coefficient and resistivity at low temperature. In the family of Hall curves for any one impurity the Hall maximum, which indicates the onset of impurity conduction, moves to higher temperature and lower values as the impurity concentration of the samples is increased.[1] In the region of impurity conduction, the resistivity curves have slopes corresponding to activation energies about an order of magnitude less than that for impurity ionization energies. The curves for samples with 10^{18} cm^{-3} < N_{maj} < 5×10^{18} cm^{-3} are of rather complex nature and exhibit more than one resistivity slope and often more than one Hall maximum.

2. Discussion of the Separate Ranges of Impurity Conduction

The breakdown of the phenomenon into separate ranges can best be illustrated by the plot of resistivity at 14·2°K against N_{maj} for many B doped samples in Fig. 1. This curve is similar to one used by Fritzsche[2] and discussed at length by James[3] for germanium.

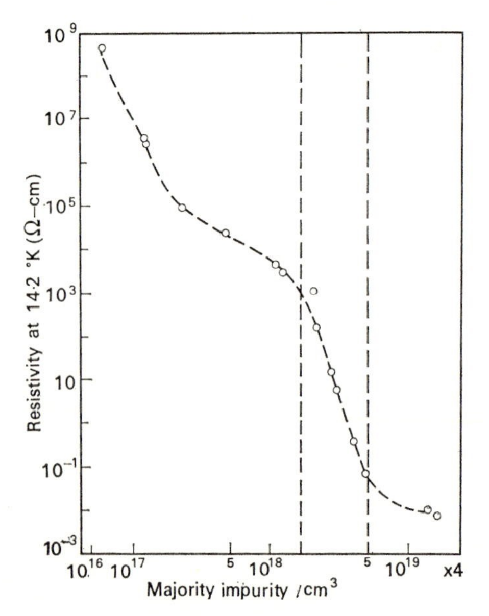

Fig. 1. The resistivity of B doped silicon at 14·2°K vs. the majority impurity concentration.

For $N_{maj} < 10^{18}$ cm^{-3} the conduction process is believed to be one of charge permutation between randomly distributed impurities. This can only occur if at least a small amount of compensation exists for the purpose of providing empty majority sites. Additional compensation should enhance the effect in this range. According to Mott[4] when $N_{maj} \gg N_{min}$ the activation energy for this type of impurity conduction should be of the order $\epsilon = 2e^2k^{-1} N_{maj}^{\frac{1}{3}}$. The activation energies, taken from the low temperature slopes of sample resistivities in the low impurity concentration range, increase with increasing impurity concentration approximately in the manner indicated by Mott.

For samples with $N_{maj} > 5 \times 10^{18}$ cm^{-3} the Hall mobilities are higher than for samples in the low concentration range. This conduction is believed to take place in a band within the forbidden gap which is made up of randomly distributed impurity states which are close enough together such that the wave functions overlap. The conduction in such a band would then be similar to metallic conduction in a half-filled band. For such a case compensation should decrease the number of carriers participating, without improving the mobility of these carriers and thereby decrease the effect.

The sharp drop in resistivity between the high and low concentration range has been termed the transition range by James, who reasons that in going from the isolated impurity case to the overlapping wave function condition there may be a transition range in which conduction may take place along preferred paths of impurities.

3. Introduction of Impurity Compensation

Irradiation with 10 MeV deuterons was used to introduce controlled amounts of compensation in B and Sb doped samples. Previous knowledge[5] of the location of the irradiation introduced defect states made it possible to tailor the experiments in such a way that the defect states did not influence the observation of the chemical impurity conduction, other than the effect desired

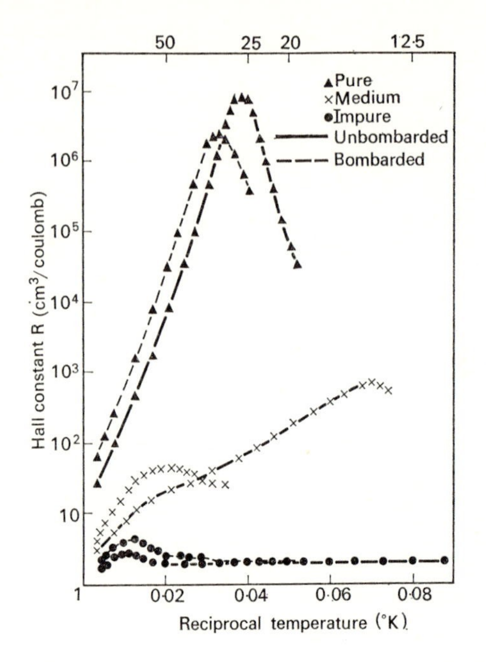

FIG. 2. The effect of compensation by deuteron irradiation on the Hall coefficient of B doped silicon samples in the impurity conduction temperature range.

due to compensation. The Hall coefficient curves of Fig. 2 show the chemical impurity conduction for three B doped samples from the low, intermediate and high impurity ranges before and after introducing about 15–25% compensation by irradiation.

For the pure sample, the manner in which the process is enhanced by compensation is strikingly evidenced by the shift of the Hall maximum to higher temperature and lower value. The corresponding resistivity curves showed a decrease in the activation energy characterizing the process from $5 \cdot 34 \times 10^{-3}$ eV to $3 \cdot 4 \times 10^{-3}$ eV. This amount of decrease agreed rather well with that predicted by the approximate relation, after Koenig and

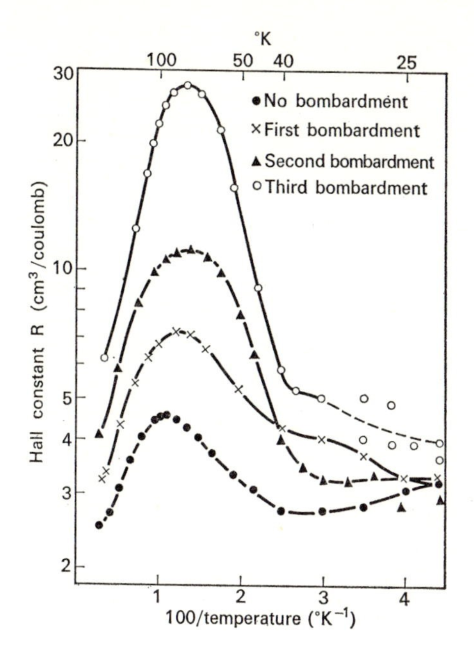

FIG. 3. Hall coefficient of near degenerate B doped silicon in the impurity conduction range after successive deuteron irradiations.

Gunther-Mohr,[6] for the activation energy in samples containing appreciable compensation $\epsilon \simeq 0 \cdot 62 e^2 k^{-1}(N_{maj}^{\frac{1}{3}} - N_{min}^{\frac{1}{3}})$.

The results from the most impure sample indicate that the impurity conduction has decreased considerably at intermediate temperatures but is relatively unaffected at temperatures below about 25°K. The corresponding resistivity curve however shifted upward over the whole temperature range. The Hall coefficient results for the same sample are shown again in Fig. 3 before and after three successive irradiations. It is not clear why the curve after the second irradiation crossed over that for the first irradiation in the vicinity of 40°K. However all four curves come approximately together below about 25°K (not shown in Fig. 3).

The Fermi level remains between the acceptor state introduced by deuteron irradiation and the valence band for every stage of irradiation. Therefore it is possible that the hole filled defect acceptor states have formed a second band which manifests itself below about 25°K and the sum of the contributions from the partially filled chemical band and the filled defect band is approximately equal to the contribution from the filled chemical impurity band before compensation.

Still another interesting result in Fig. 3 is the increase in the B impurity ionization energy from 0·0095 eV to 0·017 eV after successive irradiations. This is due, in part at least, to the removal of screening of the impurity states by free carriers.

Referring back to Fig. 2, the intermediate sample has been selected from the transition range. After the first maximum the Hall coefficient curves rise again; a second maximum sometimes follows. The resistivity curves for such samples can be characterized by three slopes in the manner indicated by Fritzsche.[2]

$$\frac{1}{\rho} = c_1 \exp(-\epsilon_1/kT) + c_2 \exp(-\epsilon_2/kT) + c_3 \exp(-\epsilon_3/kT)$$

where ϵ_1 is the usual impurity ionization energy, ϵ_3 is the activation energy for impurity conduction found in the low concentration samples and ϵ_2 is a new intermediate activation energy that is found for samples in the transition range. The compensation introduced for this intermediate sample has decreased the impurity conduction in the overlap region but enhanced it considerably at lower temperatures. This suggests that perhaps intermediate range samples show conduction by an excited state band in the higher temperature regions[7,8] followed by freezeout to the ground state and then conduction through the carrier jumping process for the ground states in the lowest temperature regions.

REFERENCES

1. R. K. Ray, T. A. Longo and K. Lark-Horovitz, *Bull. Amer. Phys. Soc.* II, 3, 101 (1958).

2. H. FRITZSCHE, *U.S. Signal Corps Quart. Rep.* Purdue University, Oct–Dec. (1956).
3. H. M. JAMES, *U.S. Signal Corps Quart. Rep.* Purdue University, Oct–Dec. (1956).
4. N. F. MOTT, *Canad. J. Phys.* **34,** 1356 (1956).
5. T. A. LONGO and K. LARK-HOROVITZ, *Bull Amer. Phys. Soc.* II, **2,** 157 (1957).
6. S. H. KOENIG and G. R. GUNTHER-MOHR, *J. Phys. Chem. Solids*, **2,** 268 (1957).
7. G. A. SWARTZ, *Bull. Amer. Phys. Soc.* II, **2,** 134 (1957).
8. S. H. KOENIG and G. R. GUNTHER-MOHR, see ref. 6, p. 278.

13

Deuteron and Neutron Bombardment of Semiconductors (abstracts of first experimental work in this field)†

Deuteron-bombarded Semiconductors‡

Pure germanium (resistivity of $\sim$ 1 to 5Ω-cm) and *P*- and *N*-type germanium semiconductors have been bombarded with $\sim$ 10-MeV deuterons. The resistivity of the pure samples decreases after bombardment. The resistivity of the *P*-type samples, measured both as bulk resistivity and as spreading resistance, also decreases. *N*-type samples show a tendency to a transformation into *P*-type, and the resistance, measured as spreading resistance, increases. These effects, observed immediately after bombardment, remain after the radioactivity of the sample has disappeared. Rectification and photo-effect change after bombardment with the change in sign of the carriers. From the measured radioactivity of the bombarded sample, the deuteron current and the number of recoils in the sample calculated from Coulomb scattering, the effect of new impurity centers, trapped hydrogen atoms and lattice defects produced by recoil can be estimated.

Neutron-bombarded Germanium Semiconductors§

It has been shown recently[1] that in germanium semiconductors bombarded by deuterons and alpha-particles lattice defects are

† *Phys. Rev.* **73**, 1256 (1948), and *ibid.* **74**, 1255 (1948), with E. Bleuler, R. E. Davis, W. E. Johnson, S. Siegel, and D. J. Tendam.

‡ Assisted by Signal Corps and ONR contracts.

§ The work at Purdue was supported by a Signal Corps Contract.

produced which act as acceptors, thus producing "hole" conduction (*P*-type) (as tested by direction of rectification, Hall effect, and photo-effect). *P*- and *N*-type germanium samples with known electrical characteristics have been irradiated with neutrons in the Oak Ridge reactor; resistivities in *N*-type semiconductors increase under neutron irradiation until the donators are neutralized, and the resistance reaches a maximum; under continued irradiation the conductivity increases again. In *P*-type germanium the conductivity increases steadily with neutron irradiation and values 100 to 200 times the original conductivity have been reached. Whereas the deuteron induced lattice defects are healed by prolonged heat treatment at 400°C, only part of the induced neutron defects are healed in a comparable time interval. Experiments with commercial *N*-type germanium rectifiers indicate changes in the forward and backward direction in agreement with the conclusion that the *N*-type germanium has been converted to *P*-type. The effect seems more pronounced in Cd-shielded samples.

Reference

1. K. Lark-Horovitz, E. Bleuler, R. Davis, and D. Tendam, *Bull. Am. Phys. Soc.* **23,** No. 3, 25 (1948).

14

Neutron Irradiation of Semiconductors†

Neutron Irradiated Semiconductors

Germanium semiconductors after bombardment with deuterons or alpha-particles show[1] permanent changes in their electrical properties: the resistivity of *P*-type germanium decreases, indicating the production of acceptors[2] due to the bombardment, and *N*-type germanium becomes converted to *P*-type. The original state can be reproduced by heat treatment, indicating that the observed effects are not due to transmutations, but primarily due to lattice displacements, creating vacancies and interstitial atoms. If this explanation is correct, it is expected to find similar changes due to neutron bombardment.

Various samples of Ge semiconductors, with known impurity type and content[2] were exposed in the Oak Ridge reactor and the conductivity was measured during exposure as a function of irradiation: *P*-type samples show an increase in conductivity, as expected from the results of cyclotron irradiation (Fig. 1*a*).

N-type material shows a decrease in conductivity which reaches a minimum, and then increases steadily with continuing bombardment (Fig. 1*b*). Hall effect measurements[2] after bombardment indicate that the material has been converted to *P*-type. These experiments indicate also that neutron bombardment produces acceptors, creating additional holes; thus the conductivity of *P*-type material increases whereas the conductivity of *N*-type material first decreases, as electrons are removed from the conduc-

† *Phys. Rev.* **76,** 442 (1949), with W. E. Johnson; *ibid*; **78,** 814 (1950) with J. W. Cleland and J. C. Pigg; *ibid.* **78,** 815 (1950), with J. H. Crawford, Jr.; and *ibid.* **79,** 889 (1950), with J. H. Crawford, Jr.

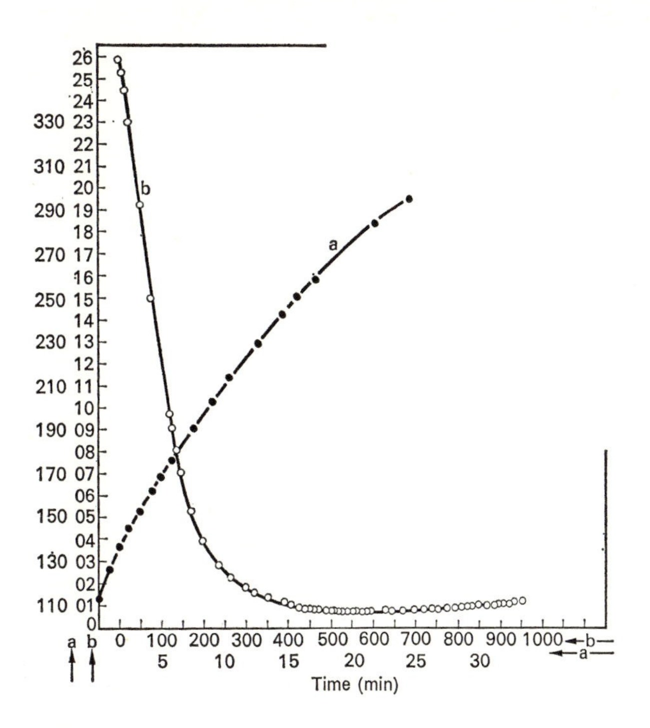

FIG. 1. Curve *a:* conductance of high resistance *P*-type germanium as a function of the time or irradiation. Curve *b:* conductance of *N*-type germanium as a function of time of irradiation.

tion band, but increases again as hole conduction becomes prominent.

Exposure of germanium point contact rectifiers of the 1N34 and 1N38 type shows[3] that at 3 volt forward bias, where spreading resistance is primarily responsible for the resistance measured, the behavior is the same as the one observed for bulk resistivity (Fig. 2*a*). At forward bias of less than 0·18 V and for bias in the back direction the resistance is determined primarily by the barrier resistance of the rectifier and since this barrier in an *N*-type

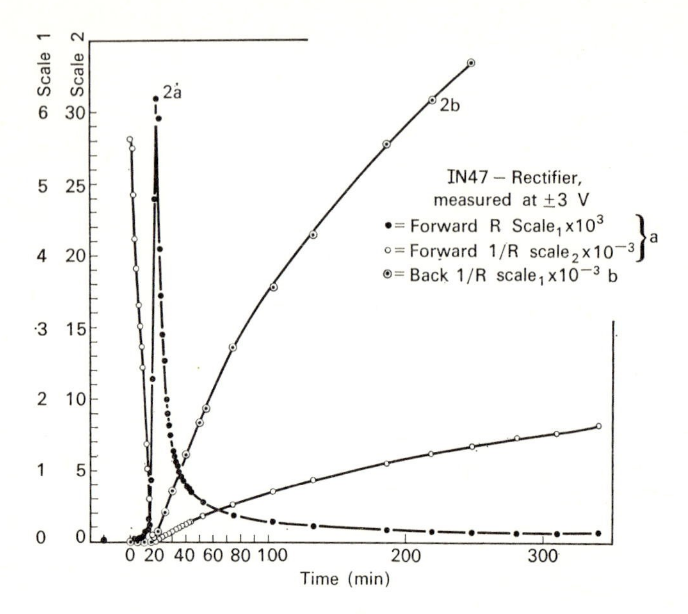

FIG. 2. Resistivity behavior in the forward and backward direction of a contact rectifier as a function of the time of irradiation.

rectifier transparent for holes a decrease of resistance is expected and is actually observed (Fig. 2*b*).

The rectifier shows after prolonged irradiation ohmic behavior, but when reassembled with an aluminum point shows the behavior of a *P*-type rectifier.

When exposed in cadmium shields, the samples reach the point of minimum conductance at larger nvt values than in a graphite torpedo. Heat treatment reproduces the original behavior observed before neutron irradiation as in the case of deuteron and alpha-bombardment.

Silicon samples both *N*- and *P*-type show an increase in resistivity in a similar way, as has been observed with deuteron bombarded silicon samples. Heat treatment restores the original resistivity values; it also shows that deep lying traps of about

0·5 eV above the full band are produced, removing electrons and holes, thus producing poorly conducting material.

Preliminary experiments on Si-rectifiers (IN 21) Cu_2O semiconductors in bulk and rectifiers, Se bulk material and rectifiers, all show a behavior similar to the one observed for silicon samples discussed above.

Transmutation-produced Germanium Semiconductors

Germanium exposed in a nuclear reactor becomes a *P*-type semiconductor ("hole" or defect conduction).[4] It has been shown that this change is due to (a) lattice displacements produced by fast neutrons and (b) impurity centers produced by the transmutations due to slow neutrons. That this latter defect must exist became apparent when heat treatment at 450°C failed to restore the original conductivity in samples which had been exposed for a long period ($nvt \sim 10^{18}$). The cross sections for activation given in the literature[5] lead to an *N*-type semiconductor and therefore K. Lark-Horovitz suggested a re-determination of the capture cross section on separated germanium isotopes. Such a determination was carried out by H. L. Pomerance,[6] and is shown in Table 1.

The capture cross sections obtained by Pomerance predict an excess of *P*-type impurities (Ga) over *N*-type impurities (As).

$$N_e = nvt \times \sigma_e \times N_{Ge} \times P_i, \tag{1}$$

where N_e is the number of events observed, nvt is the integrated neutron flux, σ_e is the cross-section for the event, N_{Ge} is the number of atoms per cc in the material and P_i is the isotopic percentage of the atomic species. Therefore,

$$N_{Ga} - N_{As} = nvt \times 4{\cdot}4 \times 10^{22}(3{\cdot}25 \times 21{\cdot}2 - 0{\cdot}6 \times 37{\cdot}6)10^{-26}$$
$$= 2{\cdot}05 \times 10^{-2} nvt. \tag{2}$$

The impurity centers are formed at the lattice sites themselves. The semiconductor thus obtained, if one starts with "pure" single crystals of germanium, is therefore a substitutional alloy.

TABLE 1. ABSOLUTE CROSS SECTION OF SEPARATED GERMANIUM ISOTOPES

(L. Seren, Argonne Laboratory)				
Isotope	Abundance	Capture cross-section (in barns)		End product
		Isotopic	Atomic	
Ge^{70}	21·2	0·073	0·0155	Ga
Ge^{70}	21·2	0·45	0·095	Ga
Ge^{74}	37·1	0·38	0·14	As
Ge^{76}	6·5	0·085	0·0055	Se
(H. L. Pomerance, Oak Ridge Laboratory)				
Isotope	Abundance	Capture cross-section (in barns)		End product
		Isotopic	Atomic	
Ge^{70}	21·2	3·25	0·69	Ga
Ge^{72}	27·3	0·94	0·26	Ge
Ge^{73}	7·9	13·69	1·08	Ge
Ge^{74}	37·6	0·60	0·22	As
Ge^{76}	6·1	0·35	0·02	Se

Germanium samples with known type and number of impurity centers (as determined by the Hall effect and the dissociation equation)[7] are exposed for various lengths of time in the reactor, in a region of determined flux. After exposure, Hall effect and resistivity are determined giving *P*-type centers due to (a) displacements, (b) transmutations. The sample is then heat treated for various lengths of time at 450°C until further heat treatment no longer produces any changes. Provided the rate of cooling from 450°C does not freeze in any lattice defects, and that all displacements produced by knock-on collisions are completely annealed out at this low temperature, the resulting conductivity is now due to the originally introduced chemical impurities and the newly produced impurities from transmutations. The experimental results are shown in Figs. 1 and 2.

The Hall curve obtained in a sample containing lattice defects

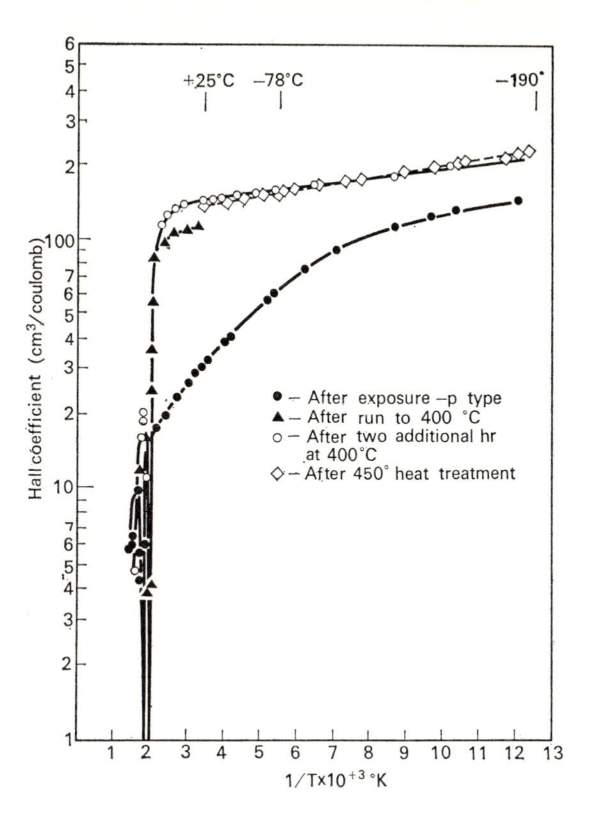

FIG. 1. Hall coefficient vs. temperature.

cannot be interpreted by an ordinary dissociation equation, since it shows a curvature which is usually not observed in chemically prepared samples. Quantitatively, Table 2 shows good agreement between the final number of carriers calculated and determined experimentally within the limits of experimental error (20% for flux measurements, 20% for cross sections, and heat treatment).

Activation cross-sections for deuterons measured by E. Bleuler and D. Tendam at Purdue lead to the conclusion that deuteron bombardment should lead to *N*-type germanium. Similarly α-activation should also lead to *N*-type germanium. Activation by fast neutrons (Cd shielded samples) should lead to a larger preponderance of *P*-type centers (Ga) but with a much smaller cross-section than slow neutron activation. Preliminary experiments are in agreement with this expectation.

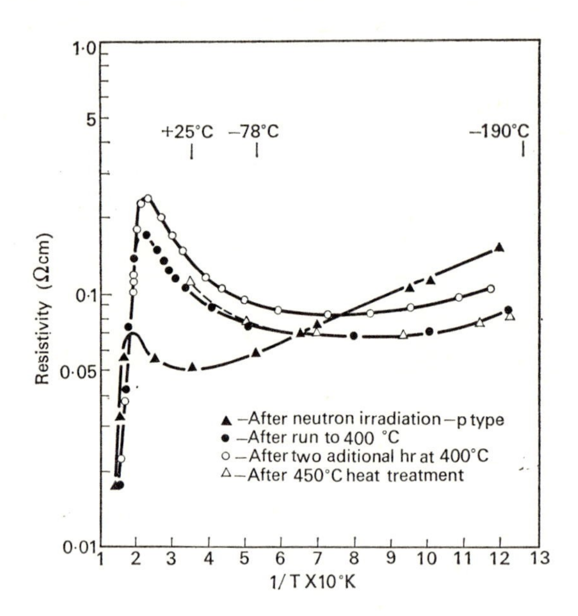

FIG. 2. Resistivity vs. temperature.

We find, therefore, (a) that there is a balance between donators and acceptors, and that the Hall effect and conductivity are determined by the excess type present, (b) that there is one current carrier released per impurity center produced, (c) agreement between the activation energies of impurity centers produced chemically and by transmutations, indicating that chemical doping is a substitution process.

TABLE 2. DENSITY OF IMPURITY ATOMS IN GERMANIUM

Sample number	Original number of carriers/cm³ (n_0)	Total flux (nvt)	Final number of carriers/cm³ (nf)		Difference (%)
			Calculated from cross-sections	Measured by Hall effect	
1—*N*-type	$-5\cdot25 \times 10^{14}$	$4\cdot56 \times 10^{17}$	$+8\cdot82 \times 10^{15}$	$+6\cdot16 \times 10^{15}$	− 27
2—*N*-type	$-2\cdot28 \times 10^{16}$	$2\cdot54 \times 10^{18}$	$+2\cdot92 \times 10^{16}$	$+3\cdot21 \times 10^{16}$	− 9
3—*N*-type	$-5\cdot47 \times 10^{16}$	$4\cdot37 \times 10^{18}$	$+3\cdot48 \times 10^{16}$	$+5\cdot25 \times 10^{16}$	+ 33
4—*N*-type	$-1\cdot17 \times 10^{16}$	$2\cdot54 \times 10^{18}$	$+4\cdot03 \times 10^{16}$	$+5\cdot20 \times 10^{16}$	+ 22
5—*N*-type	$-1\cdot48 \times 10^{14}$	$2\cdot54 \times 10^{18}$	$+5\cdot05 \times 10^{16}$	$+4\cdot45 \times 10^{16}$	− 13
6—*P*-type	$+5\cdot02 \times 10^{14}$	$2\cdot54 \times 10^{18}$	$+5\cdot15 \times 10^{16}$	$+4\cdot95 \times 10^{16}$	− 4
7—*P*-type	$+4\cdot60 \times 10^{14}$	$8\cdot04 \times 10^{17}$	$+6\cdot25 \times 10^{16}$	$+5\cdot22 \times 10^{16}$	− 20
8—*N*-type	$-4\cdot04 \times 10^{14}$	$1\cdot06 \times 10^{19}$	$+2\cdot15 \times 10^{17}$	$+1\cdot90 \times 10^{16}$	− 13

Fast Neutron Bombardment Effects in Germanium

Neutron bombardment of germanium[8] in a nuclear reactor produces impurities by transmutation (thermal neutron capture[9]) and lattice displacements caused by high energy neutrons. The contribution of transmutation introduced impurities is in general negligibly small compared to the effect of damage by fast neutrons on carrier concentration in Ge.

The initial change in conductivity during neutron bombardment of N-type Ge is ($\sigma = e\mu_e n_e + e\mu_H n_H$),

$$\frac{d\sigma}{d(nvt)_{\text{fast}}} = e\mu_e \frac{dn_e}{d(nvt)_{\text{fast}}} + en_e \frac{d\mu_e}{d(nvt)_{\text{fast}}}, \tag{1}$$

where e is the electronic charge, n_e is the electron concentration, μ_e is the electronic mobility and $(nvt)_{\text{fast}}$ is the integrated fast neutron flux. The contribution of the added scattering centers will be a negligible portion of the mobility until their concentration becomes comparable to the original impurity concentration. In the initial linear portion of the conductivity vs. bombardment curve the finite rate of annealing of neutron produced damage can be neglected, consistent with the fact that the number of acceptors per incident neutron is the same at both dry ice and room temperature. Thus, initially,

$$d\sigma/d(nvt)_{\text{fast}} = -e\mu_e K, \tag{2}$$

where K is the average net number of acceptors produced per incident neutron (K is a weighted average depending on the energy spectrum of the neutron flux in the high energy range) evaluated from the initial slopes of the conductivity vs. bombardment curve if the initial mobility is known.

Thermal equilibrium between electrons and holes when classical statistics are valid leads to[10]

$$n_e n_h = AT^3 \exp(\Delta\epsilon_g/kT) = \kappa(T), \tag{3}$$

where for Ge, $A = 5 \cdot 3 \times 10^{32}$ and $\Delta\epsilon_g = 0 \cdot 75$ eV. Thus using Eq (3) and c as the ratio of electron to hole mobility,

$$\sigma = e\mu_h[n_e c + \kappa(T)/n_e]. \tag{4}$$

The electron concentration for minimum conductivity is then

$$n_e = (\kappa(T)/c)^{\frac{1}{2}} \tag{5}$$

and the minimum conductivity is

$$\sigma_{\min} = (2e/c)\mu_e(\kappa c)^{\frac{1}{2}} \tag{6}$$

from which μ_e at the minimum can be calculated. Any photoelectric effects induced by β- and γ-radiation would cause the calculated value to be apparently larger than the expected value. The actual value should be slightly smaller than the original value of the mobility because additional scattering centers are introduced during bombardment.

Figure 1 shows the form of the conductivity vs. bombardment

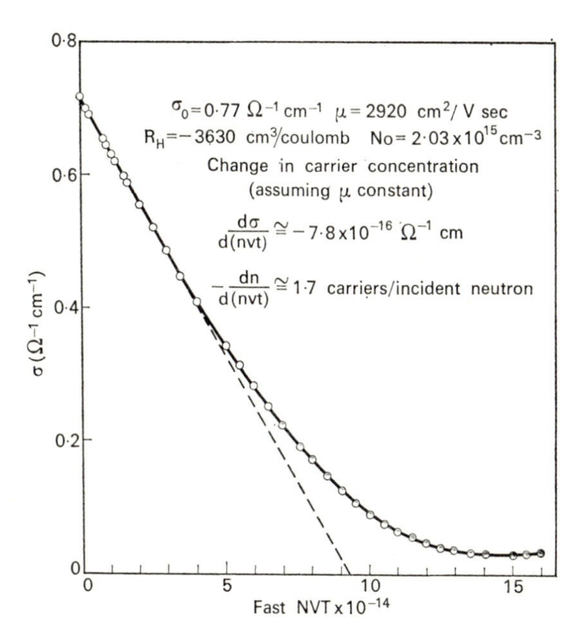

FIG. 1. Conductivity vs. bombardment for N-type germanium.

curve for 32°C. The curves taken at dry ice temperature are quite similar. Table 1 summarizes the results of the analysis.

TABLE 1. SUMMARY OF RESULTS

Sample	Temperature of exposure °C	Initial number of carriers cm^{-3}	Initial mobility cm^2/volt sec	Mobility at minimum cm^2/volt sec	Number of acceptors per incident neutron cm^{-3}
1	~ 32	$2{\cdot}0 \times 10^{15}$	2920	2200	1·7
2	30	$4{\cdot}3 \times 10^{14}$	2010	2020	1·5
3	20	$8{\cdot}9 \times 10^{14}$	2600	2190	3·5
4	~ 32	$2{\cdot}8 \times 10^{16}$	2180	1440	2·6
5	~ 32	$5{\cdot}5 \times 10^{16}$	800	13,700	5·2
6	− 79	——	10,000	——	4·3
7	− 79	——	6070	——	3·1
8	− 79	——	4530	——	3·2

The mean value of K is 3·1 acceptors per incident neutron in agreement with the theoretical value for the number of displaced atoms in Ge obtained by G. E. Evans[11] on the basis of radiation damage theory.

The difference in slope of the bombardment curve after conversion to P-type may be due to (a) the rate of recombination of lattice defects becoming appreciable after a certain concentration is reached and, (b) only those acceptors which can be thermally ionized being effective in increasing the hole concentration. Bombardment data for an initially P-type sample gives 0·77 for the number of conducting holes produced per incident neutron. This is smaller than the mean value of K given above by a factor of 4. Consequently, since recombination of defects should be negligible initially, one might suspect both of these factors to be important in the general case.

The assumed thermal equilibrium between electrons and holes seems to hold reasonably well in every case but one, sample 5 which has an abnormally low initial mobility indicating an inhomogeneous distribution of impurity centers. An impurity gradient perpendicular to the direction of current would obscure the correct minimum conductivity. Also the presence of a *P–N* barrier (the gradient parallel to current) would cause a conductivity minimum much higher than the expected value since Lark-Horovitz and co-workers[12] have shown that such a boundary is photo-sensitive to high energy radiations. In either case the calculated value of the mobility at the minimum conductivity would be spurious and high. Thus one is led to the conclusion that photoelectric effects caused by β- and γ-radiation are in general negligible in uniform samples.

Thermal Equilibrium in Neutron-irradiated Semiconductors

The conductivity vs. irradiation curves of Ge semiconductors exposed to fast and thermal neutron flux in the Oak Ridge pile were recently analyzed. The results indicate that the decrease in electron concentration in *N*-type Ge is of the order of 3 per incident fast neutron, but that the increase in hole concentration in *P*-type Ge is much smaller. Additional experiments show that, depending on temperature, 0·6 to 0·8 carrier is released initially per incident neutron. These results are tabulated in Table 1.

Bombardment effects on *N*-type and *P*-type Ge differ, since, in the case of *P*-type Ge, only those bombardment-introduced

TABLE 1. INITIAL CHANGE IN *P*-TYPE CARRIER CONCENTRATION DURING NEUTRON BOMBARDMENT

Sample (*P*-type Ge)	Temperature of exposure	Increase in hole concentration per incident neutron
1	0°C	0·61
2	20°C	0·70
3	30°C	0·77

acceptors which are thermally ionized can contribute to conductivity. This explanation holds whether these acceptors have identical or a wide distribution of activation energies.

In all experiments on *N*-type Ge carried out in the reactor and, therefore, subject to γ- and β-radiation, the maximum resistivity reached during bombardment is smaller than the predicted value, which should be even greater than that of intrinsic Ge due to the additional introduced scattering centers.

Using the mobility and equilibrium values previously reported,[13] it was concluded[14] from the expression for the minimum conductivity,

$$\sigma_{\min} = 2e\mu_e(\kappa/c)^{\frac{1}{2}} \qquad (1)$$

(μ_e the electron mobility, c the ratio of electron to hole mobility, and $\kappa = n_e n_h = AT^3 e^{-\epsilon_g/kT}$) that ionizing radiation does not disturb the thermal equilibrium, since the values of μ_e calculated from Eq. (1) at the conductivity minimum were somewhat smaller than the initial values, as was expected.

However, proper interpretation[15] of Hall effect measurements in impurity semiconductors leads to lattice mobilities which are far greater than the ones reported previously, and which agree with mobility values from drift experiments.[16] Using these new mobility values and $1 \cdot 5$ for c, the ratio of mobilities, the value of κ follows from

$$\sigma_{\text{intrinsic}} = n' e\mu_e'(1 + 1/c); \quad n'^2 = \kappa \qquad (2)$$

(n' is intrinsic concentration of electrons or holes and μ_e' the electronic lattice mobility). κ is now $3 \cdot 6 \times 10^{26}$ at 300°K as compared with the earlier value of $3 \cdot 7 \times 10^{27}$.

The revised value is considerably smaller than that formerly used, but its correctness can hardly be doubted because of the experimental data which are used for its derivation, and also, since by using these equilibrium values Johnson and Fan[17] have been able to predict correctly the temperature dependence of the energy gap in Ge. The new value is only about twice the theoretical value obtained by using the electronic rest mass

$A = 2^2(2\pi mkT)^3/h^6$ whereas the earlier value was about 20 times larger.

The new value of κ in Eq. (1) leads in all cases to apparent mobilities at the conductivity minimum far greater than the one observed at the beginning of the experiment. Therefore, in high resistance material, measured in the pile, photo-effects due to ionizing radiations do play an important part. These increase the conductivity and must be considered in the theoretical interpretation of the "in pile" conductivity measurements on Ge.

Notes

[1] Lark-Horovitz, Bleuler, Davis, and Tendam, *Phys. Rev.* **73,** 1256 (1948).

[2] We are indebted to members of the Purdue Semiconductor Laboratory: V. Bottom, J. W. Cleland, R. E. Davis, and J. C. Thornhill, for the preparation of the samples and the Hall effect measurements before and after irradiation.

[3] Davis, Johnson, Lark-Horovitz, and Siegel, *Phys. Rev.* **74,** 1255 (1948); AECD 2054 (1948).

[4] K. Lark-Horovitz, *Electrical Engineering* (December, 1949).

[5] Seren, Friedlander and Turkel, *Phys. Rev.* **72,** 888 (1947).

[6] H. L. Pomerance, AEC Report, ORNL-577 (1949).

[7] V. A. Johnson and K. Lark-Horovitz, Final Report, NDRC, Purdue (1945).

[8] Davis, Johnson, Lark-Horovitz, and Siegel, *Phys. Rev.* **74,** 1255 (1948); W. E. Johnson and K. Lark-Horovitz, *Phys. Rev.* **76,** 422 (1944); K. Lark-Horovitz and J. C. Pigg, (private communication).

[9] Cleland, Lark-Horovitz and Pigg, *Phys. Rev.* **78,** 814, 1950.

[10] V. A. Johnson (private communication).

[11] G. E. Evans (unpublished results).

[12] Orman, Fan, Goldsmith, and Lark-Horovitz, *Phys. Rev.* **78,** 646(A) (1950).

[13] Purdue Progress Report to Signal Corps, May 1, 1948, p. 12 (unpublished).

[14] J. H. Crawford and K. Lark-Horovitz, *Phys. Rev.* **78,** 815 (1950).

[15] V. A. Johnson and K. Lark-Horovitz, *Phys. Rev.* **79,** 176, 409 (1950).

[16] Pearson, Haynes, and Shockley, *Phys. Rev.* **78,** 295 (1950); Purdue Progress Report to Signal Corps, March 1, 1949, p. 11.

[17] V. A. Johnson and H. Y. Fan (unpublished).

15

Nucleon-bombarded Semiconductors†

ABSTRACT

Three types of effects due to nucleon bombardment are discussed: permanent effects due to transmutations which are particularly important in irradiation by slow neutrons; transient effects produced by the passage of heavy particles and electrons; and reversible effects produced by fast nucleons. These are due to vacancies and interstitials produced by collision, and the original condition of the lattice can be restored by heat treatment. Using fast electrons, it is possible to find the threshold energy necessary to produce a vacancy or interstitial in the germanium lattice (about 25 to 30 eV), in agreement with estimates made previously. The use of well-defined beams allows the production of *p–n* barriers in germanium and hence leads to the possibility of studying the properties of such barriers as a function of thickness and concentration of carriers. In other semiconductors, such as silicon, tellurium and lead sulphide, irradiation leads primarily to a very large decrease in conductivity.

In elementary semiconductors such as germanium and silicon, it is possible to introduce deliberately a definite number of impurity centres and to predict from the type and number of impurities introduced the number of carriers (one per impurity atom) and their sign.[1–3] It is thus possible to produce at will *n*-type (electrons) or *p*-type (holes) semiconductors of any desired conductivity within a wide range of values.

In these semiconductors it is also possible to produce predictable electrical properties by heat treatment. If germanium is heated to about 850°C and then quenched, *p*-type germanium is produced. If, however, this same material is heated for a number of hours

† *Semiconducting Materials*, p. 47, Butterworths Scientific Publications, London (1951).

at 450°C, the material becomes n-type. Heating and quenching of silicon produce primarily high-resistance material. Finally, it is possible to produce transient conductivity effects by the injection of carriers either by electron bombardment or by hole injection[4–7] through the semiconductor boundary at a metal contact. These transient effects disappear in a finite time, of the order of microsecond,[7] after the field has been removed. In heat treatment or mechanical stresses, the reaction is reversible, but the introduction of impurity centres produces permanent changes. It is possible by introducing impurity centres of the opposite type to produce a balance, such that the holes and electrons "neutralize" each other and the resulting material has a very high resistance. It is not possible to remove these centres and restore the properties of the pure material except by special chemical treatment.

All these effects can be introduced by the passage of nucleons[8–12] (deuterons, protons, neutrons, or α-particles), and some even by the passage of electrons through the material. The production of artificial radioactivity leads to end products which act as impurity centres. Therefore, if the type of reaction and its cross-section are known, it is possible to produce new types of semiconductors with predictable qualities.

Elastic collisions between the passing particles and the material produce lattice defects, vacancies and interstitials, thus introducing new energy levels and releasing carriers of either sign. These lattice defects also act as scattering centres. It also is possible that "spot heating" may take place, i.e. that small regions of the material are molten and suddenly quenched. Here not only are phenomena observed similar to those due to heat treatment, but it is also possible that, at the boundary between the solid and the molten material, surface states are produced and "frozen in".[13] Finally, the passage of medium energy electrons or γ-radiation through high-resistance material produces transient effects.[7] If the energy is high enough, then, even with electrons, enough momentum can be transferred to the target atoms to produce disorder.[14] Observations of the behaviour of semiconductors under the action of nucleons are thus of interest, because the number of carriers present in the semiconductor is so small that even short irradiations with moderate flux will produce effects which can easily be observed by changes in the electrical behaviour of the material.

SEMICONDUCTORS PRODUCED BY TRANSMUTATION

The production of new semiconductors of the permanent type by transmutations† in the material will be discussed first.

For the elementary semiconductors, such as boron, silicon, germanium, selenium, and tellurium, the total activation cross-sections for thermal neutrons are known, and it is therefore possible to estimate whether irradiation should produce appreciable effects. The mechanism of the conductivity of boron is not well understood. Tellurium and selenium have been investigated in great detail, but it is not quite clear how far impurities influence their conductivity. We have not yet been able to produce tellurium which is an *n*-type semiconductor at all temperatures.

In silicon and germanium the substitutional semiconductors are well known, but silicon has a total absorption cross-section for thermal neutrons of only $\bar{\sigma}_A = 0\cdot1 \times 10^{-24}$ cm², and the only (n,γ) process leading to an impurity centre is $Si^{30}(n,\gamma)\ Si^{31} \rightarrow P^{31}$ with Si^{30} having only 3·08% abundance. Fast neutrons can lead to Al^{27} by the process: $Si^{28}(n,2n)\ Si^{27} \rightarrow Al^{27}$. The possible transmutations in germanium are summarized in Table 1.

In germanium the total activation cross-section for thermal neutrons is $2\cdot3 \times 10^{-24}$ cm² and Pomerance[15] has recently determined the isotopic cross-section in the oxides of the separated isotopes and found the values given in Table 2. From these values it is possible to predict the production of *p*-type germanium by transmutation since the production of gallium is prevalent over the production of arsenic. From the measured cross-section it is possible to calculate the excess of acceptors (gallium atoms) over donors (arsenic atoms)

$$N_e = n_n vt \times \bar{\sigma}_e \times N_{Ge} \times P_i \qquad (1)$$

† Experiments carried out in collaboration with J. Cleland, J. C. Crawford and J. C. Pigg (O.R.N.L.).

TABLE I. SUMMARY OF NUCLEAR REACTIONS IN GERMANIUM

Isotopic abundance	Isotope			
I: MAIN REACTIONS WITH DEUTERONS AND SLOW NEUTRONS ON GERMANIUM				
%		(d,p) (n,γ)		(d,n)
21·2	Ge^{70}	$Ge^{71} \xrightarrow{11\cdot 4d} Ga^{71}$		$As^{71} \xrightarrow{50h} Ge^{71} \xrightarrow{11d} Ga^{71}$
27·3	Ge^{72}	Ge^{73}		$As^{73} \xrightarrow{76d} Ge^{73}$
7·9	Ge^{73}	Ge^{74}		$As^{74} \xrightarrow{17\cdot 5d} \{Ge^{74}, Se^{74}\}$
37·1	Ge^{74}	$Ge^{75} \xrightarrow{82m} As^{75}$		As^{75}
6·5	Ge^{76}	$Ge^{77} \xrightarrow{12h,\ 59s} As^{77} \xrightarrow{40h} Se^{77}$		$As^{77} \xrightarrow{40h} Se^{77}$
II: MAIN REACTIONS WITH FAST NEUTRONS				
%		(n,p)	$(n,2n)$	(n,α)
21·2	Ge^{70}	$Ga^{70} \xrightarrow{20m} Ge^{70}$	$Ge^{69} \xrightarrow{40h} Ga^{69}$	Zn^{67}
27·3	Ge^{72}	$Ga^{72} \xrightarrow{14\cdot 3h} Ge^{72}$	$Ge^{71} \xrightarrow{11\cdot 4d} Ga^{71}$	$Zn^{69} \xrightarrow{14h,52m} Ga^{69}$
7·9	Ge^{73}	$Ga^{73} \xrightarrow{5h} Ge^{73}$	Ge^{72}	Zn^{70}
37·1	Ge^{74}	$Ga^{74} \xrightarrow{?} Ge^{74}$	Ge^{73}	$Zn^{71} \xrightarrow{2\cdot 2m} Ga^{71}$
6·5	Ge^{76}	$Ga^{76} \xrightarrow{?} Ge^{76}$	$Ge^{75} \xrightarrow{82m} As^{75}$	$Zn^{73} \xrightarrow{< 2m} Ga^{73} \xrightarrow{5h} Ge^{73}$
III: MAIN REACTIONS WITH α-PARTICLES				
%		(α,n)	$(\alpha,2n)$	(α,p)
21·2	Ge^{70}	$Se^{73} \xrightarrow{7\cdot 1h} As^{73} \xrightarrow{76d} Ge^{73}$	$Se^{72} \xrightarrow{97d} As^{72} \xrightarrow{26h} Ge^{72}$	$As^{73} \xrightarrow{76d} Ge^{73}$
27·3	Ge^{72}	$Se^{75} \xrightarrow{127d} As^{75}$	Se^{74}	As^{75}
7·9	Ge^{73}	Se^{76}	$Se^{75} \xrightarrow{127d} As^{75}$	$As^{76} \xrightarrow{26\cdot 8h} \{Ge^{76}, Se^{76}\}$
37·1	Ge^{74}	Se^{77}	Se^{76}	$As^{77} \xrightarrow{40h} Se^{77}$
6·5	Ge^{76}	$Se^{79} \xrightarrow{?} Br^{79}$	Se^{78}	$As^{79} \xrightarrow{?} Se^{79} \xrightarrow{?} Br^{79}$

TABLE 2. ABSOLUTE CROSS-SECTION OF SEPARATED GERMANIUM ISOTOPES†

Isotope	Abundance %	Capture cross-section 10^{-24} cm²		End product
		Isotopic	Atomic	
Ge^{70}	21·2	3·25	0·69	Ga
Ge^{72}	27·3	0·94	0·26	Ge
Ge^{73}	7·9	13·69	1·08	Ge
Ge^{74}	37·1	0·60	0·22	As
Ge^{76}	6·5	0·35	0·02	Se

† After H. L. Pomerance, O.R.N.L.

where N_e is the number of events observed, $n_n vt$ is the integrated neutron flux, $\bar{\sigma}_e$ is the cross-section for the event, N_{Ge} is the number of germanium atoms per cm³, and P_i is the isotopic percentage of the atomic species. If, therefore, a germanium sample is exposed to a total flux of $n_n vt$, about 2% of this number is equal to the excess of acceptors produced per cm³ in the sample. If each impurity atom releases one carrier only (as we have assumed in our analysis of the electrical properties),[1–3] one can substantiate this prediction in the following way.[16] Germanium samples with a known number of impurity centres (either *p*- or *n*-type) are exposed for a certain time, *t*, in the nuclear reactor in a position of known thermal flux. After exposure the samples are heated to 450°C for about 24 hours (or until further heat treatment does not produce any more changes). The sample is cooled slowly to room temperature to avoid the production of lattice defects due to quenching and to anneal all the disordering effects produced by the fast neutrons (see p. 170). If purest germanium in the form of single crystals is used, then, provided all disordering effects due to fast neutrons are healed out, one produces an "ideal" semiconductor, since the newly introduced atoms are definitely located at the lattice sites. Sometimes recoil may shift these positions and produce additional disordering, but this should disappear after

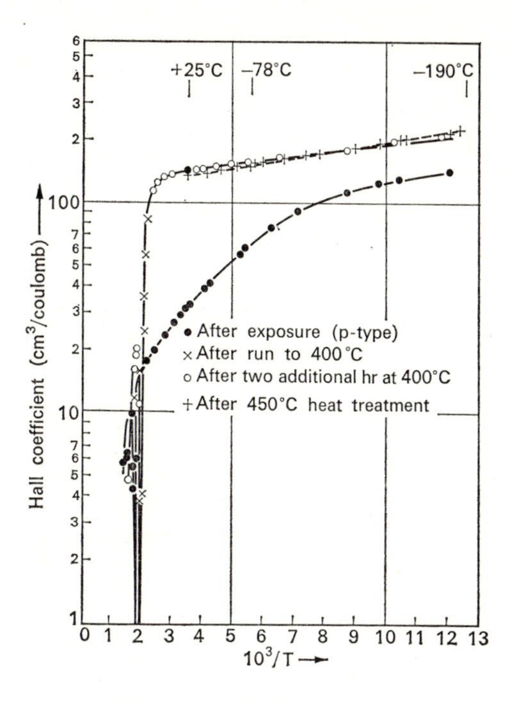

FIG. 1a. Hall coefficient as an inverse function of temperature for fast and slow neutron-bombarded *n*-type germanium. (Note effect of heat treatment: material reverts to *n*-type.)

heat treatment. The final conductivity now is due to the balance between carriers released from originally present impurities of either sign and from the impurity atoms introduced by transmutation. The accompanying figures illustrate the behaviour of Hall effect (Fig. 1a) and resistivity (Fig. 1b) after bombardment and after heat treatment. The final Hall curves are typical for an impurity semiconductor produced by substitution, whereas the Hall curves obtained immediately after exposure (slow and fast neutron irradiation) and in samples disordered by quenching show a far more complicated behaviour.

Table 3 gives a summary of the results and shows that within the limits of experimental error (both cross-sections and flux are known only within 20%) the number of carriers released and measured by Hall effect indeed equals the number of impurity centres predicted from activation cross-section and flux measurements.

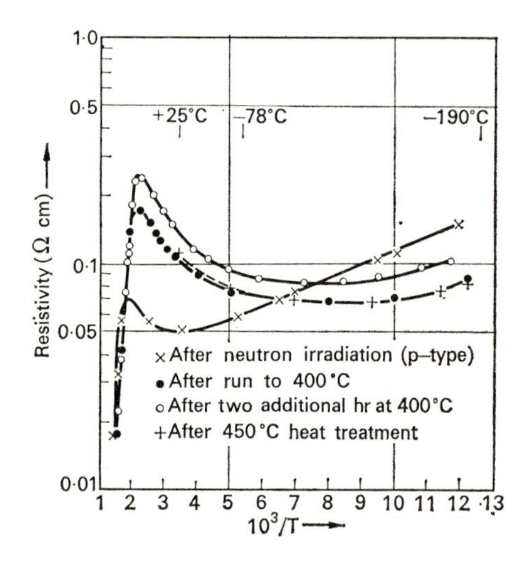

FIG. 1b. Resistivity of neutron-irradiated germanium as an inverse function of temperature. (Note the strong effect of impurity scattering before heat treatment and apparently high activation energy.)

Since semiconductors produced by transmutation are of a permanent nature and cannot be permanently changed by physical treatment, it is possible to produce high-resistance material by irradiation of originally *n*-type material in definite increments of $n_n vt$ and thus to introduce measured quantities of excess *p*-type to "neutralization". Such a semiconductor should have a conductivity somewhat smaller than an intrinsic semiconductor (because of the presence of additional scattering centres and thus a shorter mean free path) but with the same slope if the logarithm of the

TABLE 3. CONCENTRATION OF TRANSMUTATION IMPURITY ATOMS IN GERMANIUM

Sample No.	Original No. of carriers per cm³ (n_0)	Total flux ($n_n vt$)	Final No. of carriers/cm³ (n_f)		Percentage difference
			Calculated from cross sections	Measured by Hall effect	
1 (n-type)	$-5 \cdot 25 \times 10^{14}$	$4 \cdot 56 \times 10^{17}$	$+8 \cdot 82 \times 10^{15}$	$+6 \cdot 16 \times 10^{15}$	-27
2 (n-type)	$-2 \cdot 28 \times 10^{16}$	$2 \cdot 54 \times 10^{18}$	$+2 \cdot 92 \times 10^{16}$	$+3 \cdot 21 \times 10^{16}$	$+9$
3 (n-type)	$-5 \cdot 47 \times 10^{16}$	$4 \cdot 37 \times 10^{18}$	$+3 \cdot 48 \times 10^{16}$	$+5 \cdot 25 \times 10^{16}$	$+33$
4 (n-type)	$-1 \cdot 17 \times 10^{16}$	$2 \cdot 54 \times 10^{18}$	$+4 \cdot 03 \times 10^{16}$	$+5 \cdot 20 \times 10^{16}$	$+22$
5 (n-type)	$-1 \cdot 48 \times 10^{14}$	$2 \cdot 54 \times 10^{18}$	$+5 \cdot 05 \times 10^{16}$	$+4 \cdot 45 \times 10^{16}$	-13
6 (p-type)	$+5 \cdot 02 \times 10^{14}$	$2 \cdot 54 \times 10^{18}$	$+5 \cdot 15 \times 10^{16}$	$+4 \cdot 95 \times 10^{16}$	-4
7 (p-type)	$+4 \cdot 60 \times 10^{14}$	$8 \cdot 04 \times 10^{17}$	$+6 \cdot 25 \times 10^{16}$	$+5 \cdot 22 \times 10^{16}$	-20
8 (n-type)	$-4 \cdot 04 \times 10^{14}$	$1 \cdot 06 \times 10^{19}$	$+2 \cdot 15 \times 10^{17}$	$+1 \cdot 90 \times 10^{16}$	-13

resistivity is plotted against $1/T$ (Fig. 2). Figure 3 illustrates the corresponding behaviour[17] of the Hall coefficient.

These newly introduced donors and acceptors are equally effective. The conductivity/irradiation curves should have slopes differing by the ratio of mobilities ($\sigma_e = en_e b_e$, $\sigma_h = en_h b_h$ where e is the charge, n_e, n_h are the numbers of electrons and holes and

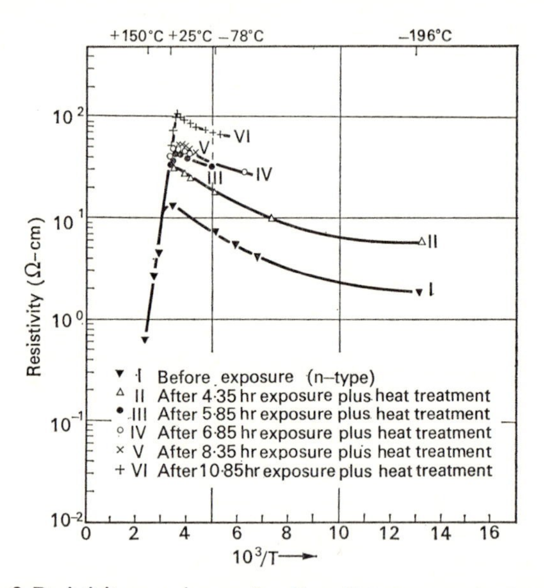

FIG. 2. Resistivity as an inverse function of temperature in neutron-irradiated germanium with successively increasing number of transmutations. (Note that resistivity is still increasing while Hall effect has decreased, cf. Fig. 3.)

b_e, b_h are the mobilities of electrons and holes). Since the newly introduced impurity atoms are permanent, no healing effects have to be considered (see in contrast pp. 170–80).

With fast neutrons, one usually produces impurities with lower atomic numbers. For example, the (n,p) reaction leads to gallium decaying to germanium, (n,α) leads to zinc decaying partially to gallium and germanium, $(n,2n)$ produces germanium decaying partially to gallium, and only the low abundance germanium 76 leads to arsenic. Therefore, if slow neutrons are excluded with cadmium, one would obtain a more pronounced shift in the direction of gallium impurities.

Deuteron irradiation. Since the effect of nucleon irradiation on the conductivity of germanium was first noticed after deuteron bombardment, it was important to estimate how far this effect

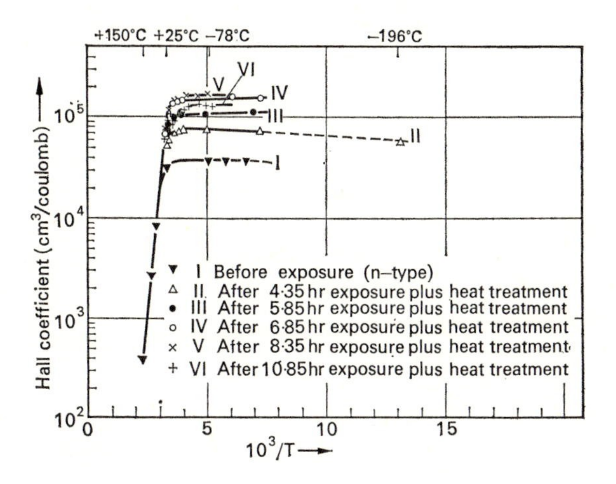

FIG. 3. Hall coefficient as an inverse function of temperature in *n*-type germanium "neutralized" by successively increasing transmutations. Maximum value reached agrees with theory (cf. Appendix I and Fig. 10).

could be due to the production of impurities by transmutation. The number of disintegrations N_d produced by any one deuteron (energy 10 MeV) can be estimated[19] approximately as

$$N_d = \bar{\sigma}_A \, r_{\text{eff}} \, N_{\text{Ge}} \tag{2}$$

where $\bar{\sigma}_A$, the activation cross-section, is in the neighbourhood[18] of 5×10^{-25} cm². In this formula, r_{eff} is an effective range taking into account the decreasing cross-section at lower energies (about 0·2 mm) or equivalent to about 0·1 g/cm², and N_{Ge} is again the number of germanium atoms per gram. With these assumptions, one finds that the number produced is of the order of 5×10^{-4} per deuteron, smaller by several orders of magnitude than the actual number of carriers determined from the conductivity change produced by deuteron bombardment which is of the order of 1 to 2.

Bleuler and Tendam[19] actually measured the activities produced by irradiating a germanium sample and, by following the decay for about three weeks, found six periods. With the proper correction factor (for absorption of β-radiation in the sample), it was possible to estimate the cross-sections, assuming (1) that the K-capture/β^+ ratio follows Fermi's theory, (2) that isomers are produced with equal probability, and (3) that the (d,n) and (d,p) (Oppenheimer–Phillips process) cross-sections are equal.

Under these assumptions it was possible to estimate the cross-sections for the various processes, and it turns out that, according to these observations, one deuteron produces $8 \cdot 3 \times 10^{-5}$ reactions: Ge^{74} (d,p) Ge^{75}. With the isotopic dilution of $37 \cdot 1\%$, this corresponds to a cross-section of $2 \cdot 3 \times 10^{-25}$ cm^2. Assuming a relative cross-section of $2 \cdot 5$ for the process Ge^{74} (d,n) As^{75} which yields no activity, the following values for the total number of new nuclei produced by one deuteron are obtained: (*a*) Immediately after bombardment the only impurity is arsenic produced by the (d,n) process, and the number of these arsenic nuclei is equal to 5×10^{-4} per deuteron; (*b*) after the decay of all radioactive nuclei, impurities of gallium, arsenic, and selenium are present as follows: $N_{Ga} = 2 \times 10^{-4}$, $N_{As} = 2 \cdot 9 \times 10^{-4}$, and $N_{Se} = 0 \cdot 27 \times 10^{-4}$, giving a total production of $5 \cdot 2 \times 10^{-4}$ new impurity centres per deuteron. A re-evaluation of the experiments, using new data on the activities involved, yields a somewhat larger amount of selenium produced. The resulting amount of arsenic is again larger than that of gallium. This may be expected without detailed calculation, since gallium results only from the (d,p) process with Ge^{70}, while arsenic is the end product of both (d,p) and (d,n) processes with the more abundant Ge^{74}.

While these numbers are subject to some uncertainty due to the lack of information on the detailed decay schemes, they are in good agreement with the estimate from theory as mentioned above. It is clear, therefore, that rather intense deuteron beams are necessary to produce effective transmutations with deuterons, but if this can be done then the material will be expected to be *n*-type because of the excess arsenic donors and not, as with the irradiation by neuterons, *p*-type. It is also possible to expose such materials inside the cyclotron, because even if melting should occur the number of transmutations produced in this way would be unchanged.

One sees from the analysis of the (α,n) and (α,p) reactions that only arsenic and selenium are produced. We have not been able by chemical addition of selenium to germanium to find any definite indication of any change in the number of carriers in germanium, and one would thus conclude that irradiation with α-particles will lead to *n*-type material through the production of arsenic. One can therefore assume that, whereas irradiation with neutrons produces *p*-type material, irradiation with charged particles, such as deuterons and α-particles, may lead to production of *n*-type material.

Lattice Defects Produced by Fast Neutrons and Charged Particles†

As mentioned in the discussion of activation by deuterons, the effects observed after irradiation with deuterons are so large that they cannot be explained on the basis of transmutations. One therefore has to assume that another effect is taking place and for this purpose it is necessary to review briefly the different processes which are observed when a particle passes through target material.

There are two problems to be considered in this discussion. The first one is the interaction of the particle with the target material,[20] which is a problem of irradiation physics[21, 22] and will concern us

† Experiments carried out in collaboration with R. E. Davis, J. Forster (Purdue), J. Cleland, J. H. Crawford, W. E. Johnson and J. C. Pigg (O.R.N.L.).

only in so far as this information is necessary for the understanding of the solid state problem. The problem which is of primary interest is the question what kind of new energy states, if any, are produced by irradiation with charged and uncharged particles.

The effects of nuclear radiations vary with the type of particle used. With γ-rays primarily transient effects, such as the photoelectric effect, the Compton effect, and pair production are observed. With a singly charged particle, the main processes to be considered are ionization and excitation. The stopping power or the energy loss for this process is well known and can be expressed by

$$-\left(\frac{dW_p}{dx}\right)_{\text{ion}} = 2\pi Z N_s \frac{r_0^2 M_p (mc^2)^2}{m W_p} \cdot \log \frac{2mv^2}{W_1} \tag{3}$$

In this formula, N_s is the number of stationary atoms/cm^3, Z the number of electrons which can be released by ionization, m the electronic mass, M_p the mass, and W_p the energy and v the velocity of the incoming particle. W_1 is a mean ionization energy of the atom and r_0 is the electronic radius. A charged particle also loses energy due to nuclear collisions

$$-\left(\frac{dW_p}{dx}\right)_{\text{coll}} = \pi N_s Z^2 r_0^2 \frac{M_p}{M_A} \frac{(mc^2)^2}{W_p} \log \left[\frac{\Delta W_p\,(\text{max})}{\Delta W_p\,(\text{min})}\right] \tag{4}$$

and here again one has Z as the charge of the stationary nucleus and M_A as the mass of the stationary nucleus. The maximum energy loss, ΔW_p (max), will be determined by the maximum energy transfer in a head-on collision; ΔW_p (min), the minimum energy loss, by the screening limit. With neutrons (mass $= M_n$), the energy loss can be written

$$\Delta W_p = W_{p0} \frac{4 M_n M_A}{(M_A + M_n)^2} \sin^2 \theta/2 = W_{p0} \frac{2 A_w}{(A_w + 1)^2} (1 - \cos \theta) \tag{5}$$

where θ is the angle of deviation in the "centre of mass system", and M_A, M_n are replaced by the atomic weight A_w and 1. The average loss of energy per collision can be written

$$\overline{\Delta W_p} = W_{p0}\, 2A_w/(1 + A_w)^2 \tag{6}$$

Therefore it is clear that neutrons and heavy charged particles are able to produce lattice defects by collision: vacancies and interstitials which profoundly alter the energy band structure of the material.

It is possible to write the elastic scattering cross-section of deuterons of energy W_p as $\bar{\sigma}(W_p, W_r^0)$, and to give an expression for the scattering cross-section with a given minimum recoil energy W_r^0

$$\bar{\sigma}\,(W_p, W_r^0) = \pi Z^2 r_0^2 \frac{M_D}{M_A} \frac{(mc^2)^2}{W_p\, W_r^0} \left(1 - \frac{M_A\, W_r^0}{4M_D\, W_p}\right) \tag{7}$$

Here M_D is the mass of the deuteron, and W_r^0 the minimum recoil energy necessary to displace a target atom, and if $\dfrac{M_A\, W_r^0}{4M_D\, W_p}$ is neglected as compared to unity, this becomes for germanium:

$$\bar{\sigma}\,(W_r^0) = \frac{1{\cdot}83 \times 10^{-18}}{W_p \times W_r^0}\ \text{cm}^2 \tag{8}$$

where W_p is expressed in MeV and W_r^0 in eV.

The range/energy relation for deuterons between 2 and 10 MeV in germanium was determined experimentally by Bleuler, Heller and Tendam, and can be expressed as $W_p = 0{\cdot}480\ r_{\text{eff}}^{0{\cdot}614}$ where W_p is in MeV and r_{eff} in g/cm². Below 2 MeV the range is somewhat higher than that given by this formula. From the range/energy relation and this cross-section one calculates the total number of primary recoils made in the sample. Assuming a minimum recoil energy necessary to displace an atom to be equal to 30 eV (see pp. 183–9) one obtains, as the cross-section, $0{\cdot}61 \times 10^{-19}/W_p$. Using this value in the range/energy relation one obtains a total number of primary recoil atoms equal to about 25, for deuterons of 10 MeV. Each primary recoil may make secondary recoils, losing its energy in this way until the momentum transfer is too small to produce a displacement. If one considers the knocked-on germanium atom like a light fission fragment, using similar range relations at lower energy, then one can estimate that the average number of secondaries is of the order of 8. This would lead to a total number of recoils of about 200. On the other hand, Seitz[21,22] has expanded his original calculations of collision losses recently to include germanium and silicon. By using his method one obtains about 175 for the total number of knock-ons produced with 10 MeV deuterons, a value which is in fair agreement with our estimate.

The actual number of carriers produced per incident deuteron is far smaller, only 1 to 2, indicating a large amount of healing during the bombardment process and during the time necessary to remove the sample and start Hall effect measurements. Even in the very first experiments carried out with α-particles we had found a considerable "decay" of the conductivity, and this should be taken into account in the evaluation of the deuteron experiments.

Calculations using the cascade process for the displaced germanium atoms produced by neutrons of a certain energy distribution around 2 MeV have been carried out by Evans[23] at the Oak Ridge National Laboratory, following procedures worked out by Brown and James. Under the assumption that the lattice perturbations are due solely to displaced germanium atoms, produced by elastic collision of fast neutrons and germanium atoms, and that no other type of radiation is effective, and that the recombination, back diffusion or annealing can be neglected, Evans treats the problem of scattering in the semiconductor as scattering by individual atoms. For any given neutron energy W_p, he assumes uniform distribution of recoil energies from 0 to $W_p \times 4A_w/(1 + A_w)^2$ where A_w is the atomic weight of germanium. For each possible value of the neutron energy the number and energy of the secondary knock-ons is computed. In computing the number of secondary knock-ons only those elastic collisions were considered which transferred 25 eV or more to a knocked-on atom, it being assumed that 25 eV is the average energy required to displace the germanium atom from its lattice position. (This, as will be seen, is not far from the truth.) It is then possible to follow the dissipation of the energy of the knocked-on atoms and to estimate the total number of atoms displaced by them. The total number of atoms scattered is $1 \cdot 19 \times 10^{-4} W_p$.

For some particular energy distribution one finds about 135 scattered atoms per collision. Assuming a scattering cross-section of one barn, the actual number scattered per incident neutron in 1 cm^3 becomes about 6. This figure is in good agreement with experimental values obtained later. It should be mentioned, however, that this apparently good agreement would be very much disturbed by another choice of cross-section. Certainly the fact that other radiation is present and that healing is taking place will have to be taken into account and may change the figure appreciably.

On the other hand, one cannot exclude the possibility that the process of "spot heating", which was first suggested some time ago by Dessauer[24] to explain the interaction of high energy radiation with living matter, may play an important part. Considering this process quite naïvely, one may say that the energy which is available melts a small portion of the germanium. On quenching, this is converted from *n*-type to *p*-type material. By using the values of heat conductivity and the latent heat of melting, it is possible to estimate the amount of melting due to recoiling atoms of given energy.[25] A germanium atom recoiling from a neutron collision with energy 10^5 eV will give up its energy in melting a region about 10^{-6} cm in radius. However, this region will cool off in an extremely short time, 10^{-12} sec,[26] and therefore very rapid quenching will take place. This problem of irradiation physics, the comparison of effects of isolated imperfections with those of the disordering due to spot heating, will take a large amount of rather precise and elaborate experimentation and its discussion is beyond the scope of the present paper. In both cases, whether the disordering is produced by knock-ons and elastic collisions, or whether it is produced by spot heating, a number of vacancies and interstitials are produced. This means that the original lattice is being perturbed and as a consequence the band structure of the lattice will be changed also.

If one assumes that the vacancies and interstitials in both germanium and silicon will act like substitutions from the third and the fifth column of the periodic system, then one would conclude that the vacancy produces empty acceptor states above the full band and that the interstitial produces full donor states below the conduction band. If both of these were produced in the same way as in chemical substitution, and if they were equal in number their effect on the conductivity could not be detected because they would neutralize one another. The activation energy for impurities due to chemical substitution is so small that practically all donors and all acceptors are ionized at room temperature. However, in displacements produced in the lattice by either heat treatment or knock-ons, one will have to assume that a distribution of activation energies is present and if the donor level is depressed below the acceptor level, the donors will not be effective. Therefore, the kind of carrier obtained depends on the type of lattice perturbation and on the energy levels in the forbidden band of the new states produced by the perturbation.

The situation seems to be different in silicon and germanium. The different band width may actually produce a different behaviour. If, in germanium, one could assume that both acceptor and donor states are produced below the middle of the forbidden band then a *p*-type semiconductor would finally result. In silicon, where the band is much wider, if both donors and acceptors fall in the middle of the forbidden band, no matter whether *p*-type or *n*-type material is used as a starting material, one will finally end with high resistivity silicon. Actually, the positions of the donors and acceptors determine the final location of the Fermi level.†

Analysis of fast neutron irradiation. During irradiation with fast neutrons it has been observed that the conductivity of *p*-type material increases monotonically with irradiation, whereas the conductivity of *n*-type material first decreases, reaches a minimum, and increases again.[10-12] It has been shown that in the first part of the conductivity curve the germanium is *n*-type, but after the

† We are indebted to G. W. Lehman for a detailed discussion of some cases of shift in Fermi level as a function of neutron irradiation, Appendix II.

conductivity minimum has been passed in the second branch of the curve, the germanium is *p*-type.[12, 27] To understand this behaviour, we assume that the lattice vacancies act as acceptors, but that these acceptors have activation energies which are larger than those which are observed with chemically introduced acceptors. On the other hand, the interstitials may be considered as donors, but there are deep-lying donor levels in the neighbourhood of the full band. One would expect that acceptor states raised slightly above the full band and donor states depressed only slightly below the conduction band would cancel each other. If such states are produced by displacement, they cannot be found by electrical measurements.

The existence of such deep-lying states can be understood, since not only electrical forces, but also elastic forces are brought into play by the vacancy and the occupancy of an interstitial lattice position. One also has to consider that with germanium and silicon the removal of an atom and inserting it into an interstitial position is equivalent to the removal of more than one electron from one place (vacancy), and the addition of more than one electron in another place (interstitial) in the lattice. As a consequence one has to consider the electrical term not as hydrogen-like but rather as a many electron problem. The removal of electrons after the first involves activation energies that are far greater than in the hydrogen-like terms of the impurities, where just one electron is missing or is added to the outer shell.

In an *n*-type material, it is clear that both deep-lying and high-lying acceptors will be filled by the electrons in the conduction band. However, in a *p*-type material it is equally clear that acceptor levels can only become active when the electrons can be excited from the full band into these acceptor states. If the acceptor states are too far away from the full band, they become active only at comparatively high temperatures. In a similar way if the donor states are far below the conduction band, they cannot act except at elevated temperatures. From this point of view it is clear that the resistivity of an *n*-type germanium conductor will increase most rapidly during initial irradiation. It is also clear that after the minimum is reached, the conductivity of the newly created p-type semiconductor will increase slowly.

This has been quantitatively shown by analysis of the irradiation curves of n- and p-type semiconductors carried out recently at Oak Ridge (Crawford).[28] We assume in the initial stages of irradiation that photo-effects due to γ- and β-radiation can be neglected and that healing also is negligible. The change in mobility can be neglected also, until a large number of lattice defects has been introduced. The equation then becomes

$$\frac{d\sigma}{d(n_n vt)_{\text{fast}}} = eb_e \frac{dn_e}{d(n_n vt)_{\text{fast}}} \tag{9}$$

where n_e is the number of conduction electrons and b_e the electron mobility and σ the conductivity. If it is assumed that all the acceptor states lie far below the *initially* available donor levels, each state produced will remove an electron from the conduction band. Therefore, for the initial portion of the curve one has

$$\frac{d\sigma}{d(n_n vt)_{\text{fast}}} = eb_e\, \bar{z}_a \tag{10}$$

In this equation $\bar{z}_a$ is the average net number of effective acceptors produced by irradiation; $\bar{z}_a$ is a weighted average and depends markedly on the energy distribution of the neutrons used. The initial value of $\bar{z}_a$ may be evaluated from the initial slope if the initial mobilities are known. Table 4 shows the outcome of such an analysis for measurements at various ambient temperatures. The number of carriers per incident neutron in n-type material is about 3, in p-type material 0·6 to 0·8, dependent on the temperature. This is in agreement with the picture of energy levels assumed. The number of carriers obtained in n-type material per incident neutron at dry ice temperature is about the same. This indicates that, for the start, our assumptions are justified.

This number is also in agreement with the theoretical value obtained by Evans in his analysis of irradiation of germanium with 2 MeV neutrons. The number of 135 displacements per

TABLE 4. EXPERIMENTAL DETERMINATION OF NUMBER OF ACCEPTORS PER NEUTRON

Sample	Temperature of exposure °C	Initial number of carriers cm^{-3}	Change in number of effective carriers per incident neutron cm^{-3}
1	~ 32	$2 \cdot 0 \times 10^{15}$	$-1 \cdot 7$
2	30	$4 \cdot 3 \times 10^{14}$	$-1 \cdot 5$
3	20	$8 \cdot 9 \times 10^{14}$	$-3 \cdot 5$
4	~ 32	$2 \cdot 8 \times 10^{16}$	$-2 \cdot 6$
5	~ 32	$5 \cdot 5 \times 10^{16}$	$-5 \cdot 2$
6	− 79	—	$-4 \cdot 3$
7	− 79	—	$-3 \cdot 1$
8	− 79	—	$-3 \cdot 2$
9*	30	$1 \cdot 02 \times 10^{16}$	$-3 \cdot 9$
10	0	$2 \cdot 5 \times 10^{14}$	$+0 \cdot 60$
11	20	$4 \cdot 2 \times 10^{14}$	$+0 \cdot 70$
12†	30	$1 \cdot 6 \times 10^{15}$	$+0 \cdot 77$

* Nos. 1 to 9 all *n*-type. † Nos. 10 to 12 all *p*-type.

scattered neutron corresponds to about 6 displacements per incident neutron. Since Evans has not taken into account any healing or recombination, this is in agreement with the experimental results.

Figure 4 shows the form of the conductivity/irradiation curve for *n*-type germanium at 32°C. The curves obtained at dry ice temperature are quite similar. The initial portion is sufficiently linear to make an accurate slope determination.[28] Figure 5 shows an irradiation curve obtained for high resistivity *p*-type germanium. It is of interest to note the perfectly linear slope obtained and that changes can be measured in a few minutes, a fact which makes this experiment of value for practical purposes. Additional information about the behaviour of the irradiated semiconductor can be obtained by considering the condition for the minimum conductivity.[29] This tells us to what extent high intensity β- or γ-radiation, present in the pile, influences the behaviour at the

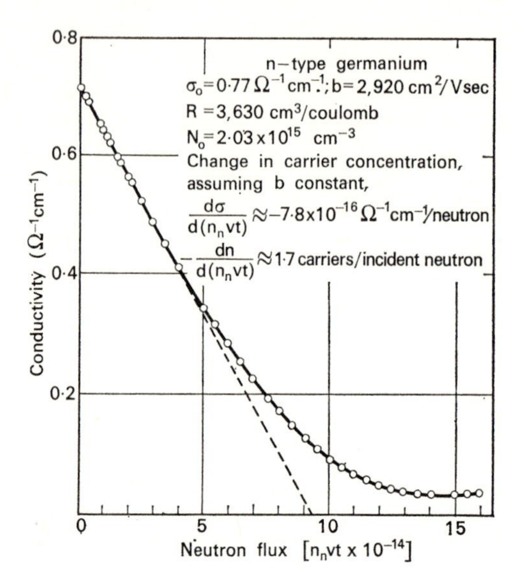

FIG. 4. Conductivity of *n*-type germanium at 32°C as a function of neutron flux.

minimum. If such transient effects take place, then we are definitely not dealing with a semiconductor in thermal equilibrium.†

The condition for thermal equilibrium between holes and electrons in the range of concentration for which classical statistics can be used can be written

$$n_e\, n_h = K(T) = \frac{4(2\pi \overline{m*} kT)^3}{h^6} \exp\,(-\,\epsilon^*/kT) \qquad (11)$$

where $\overline{m*}$ is the geometric mean of the effective masses of the carriers, ϵ^* is the width of the forbidden band and $K(T)$ is the equilibrium constant, which depends only on temperature. In former discussions we have used the value $3 \cdot 7 \times 10^{27}$ cm^{-6} for K (300°K). The use of the factor $3\pi/8$ in the Hall constant is

† See p. 179.

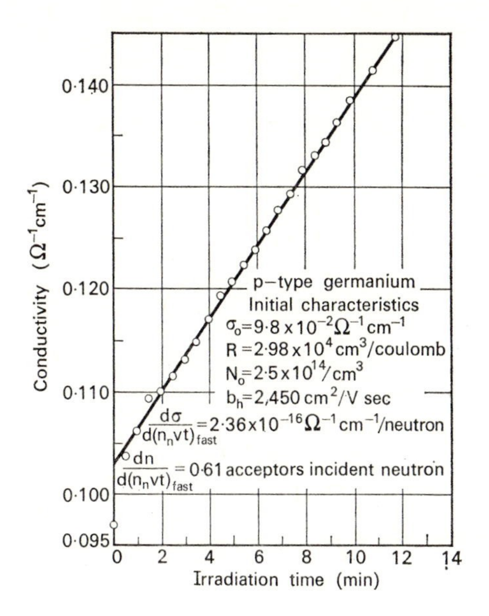

FIG. 5. Conductivity of *p*-type germanium at 0°C as a function of irradiation time (fast neutrons).

justified only if impurity scattering is negligible. This factor may vary and with it the value of the mobility calculated. However, the lattice mobility for electrons has been determined by drift measurements[310] as $b_{Le} = 3300$ cm²/volt sec. If we introduce this lattice mobility into the equation for intrinsic conductivity we find

$$\sigma_i = neb_e\,(1 + 1/\Lambda) \tag{12}$$

where n is the number of intrinsic electrons and holes. Λ, the ratio of electron to hole mobility, is 1·5 in germanium.[31] Since $K^{1/2} = n$ we now find

$$K_{300\,°\mathrm{K}} = 3 \cdot 6 \times 10^{26}\ \mathrm{cm}^{-6} \tag{13}$$

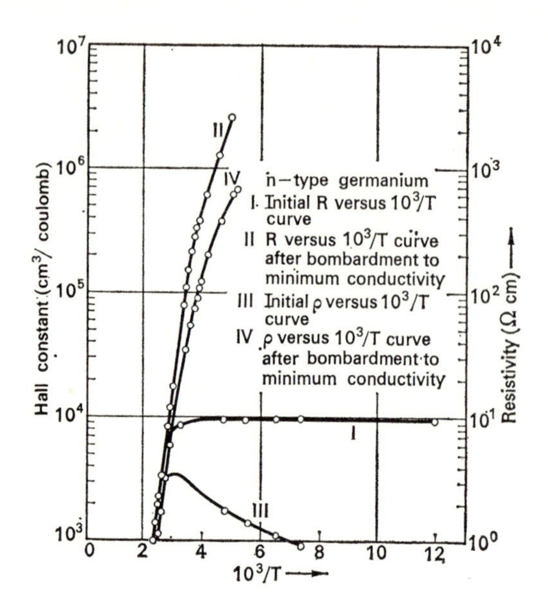

FIG. 6. Hall effect and resistivity of *n*-type germanium bombarded with fast neutrons to minimum conductivity.

So long as the ratio of electron to hole mobility is unchanged by addition of new types of scattering centres, one can write the conductivity equation as follows

$$\sigma = en_e b_e + en_h b_h = eb_h \left[n_e \Lambda + \frac{K(T)}{n_e} \right] \tag{14}$$

The electron concentration for minimum conductivity is then

$$n_e = (K/\Lambda)^{\frac{1}{2}} \tag{15}$$

The minimum conductivity is therefore

$$\sigma_{\min} = 2eb_h(K\Lambda)^{\frac{1}{2}} \tag{16}$$

These relationships and the experimental value of the minimum of conductivity allow determination of the mobilities in the semiconductor, and these in turn can be compared with the original mobility obtained from the initial Hall effect and conductivity. Transient ionization would increase the conductivity and, as a consequence, the apparent calculated electron mobility would be far larger than the original. Actually, because of the additional scattering centres introduced by bombardment, the mobility at the conductivity minimum must be smaller than at the start if the semiconductor is in thermal equilibrium.

In every case the calculated mobilities indicate that the assumption of thermal equilibrium between carriers in an *irradiated* semiconductor is not fulfilled.[29] The fact that the observed conductivity minimum is always far higher than the value of intrinsic conductivity at this temperature ($\sigma_i = 0{\cdot}0164$) can be due either to the effect of ionizing irradiation and its transient effects or to inhomogeneities in the material.

If impurity concentration gradients were perpendicular to the direction of the current one could never observe the correct minimum conductivity. The system would be equivalent to a set of parallel resistors some of which would always have conductivity higher than the minimum. On the other hand, if the concentration gradient is parallel to the direction of current, then bombardment produces *p–n* boundaries between good conducting *p*- and *n*-type semiconductors, and, since it was shown by recent experiments at Purdue that a *p–n* boundary acts as a counter (see below) and is affected by γ- and β-rays, it is clear that the expected proper minimum conductivity is not reached, and as a consequence the apparent mobilities will turn out to be very much too high if calculated from the equation for σ_{min}.

The fact that the slope of the conductivity/time of bombardment curve after conversion to *p*-type is much less than the initial slope of the original *n*-type material—less than one tenth of the initial slope—is due not only to the distribution of the acceptor states, such as have been mentioned above, but also to recombination of lattice defects, which becomes appreciable after a certain concentration is reached. The healing of the effects might be studied more easily from the decay curve after irradiation than from the irradiation curve. Such neutron experiments are under way now.† However, it is possible to predict how the conductivity should change during irradiation as a function of recombination. It appears reasonable to write the equation for the production of vacancies in the following form

$$dN_v/dt = z_d - C_r (N_v - \Delta)N_i \tag{17}$$

Here z_d represents the rate of production of defects and the second term represents the rate of recombination of interstitials, N_i, and vacancies, N_v. The effective cross-section of the vacancies being reduced by their clustering, as indicated by $(N_v - \Delta)$. Crawford's preliminary analysis[34] of the conductivity changes in *p*-type germanium irradiated with fast neutrons indicates that healing can be better represented as a first order than as a second order process. The analysis of the α-irradiation curve leads to a similar conclusion.[33]

If germanium is exposed in the reactor until minimum conductivity is reached and is then withdrawn, one would expect the slope of the log ρ versus $1/T$ curve to approach the intrinsic value. Figure 6 shows that this behaviour has been approached, but that at low temperatures the activation energy, while far higher than usually observed for germanium, is still below the intrinsic value (0·31 eV instead of 0·75 eV). This result should be compared with the results of the transmutation experiment, where acceptors and donors are equally efficient.[35]

In other materials the effects are strikingly different from those observed in germanium. The most interesting case is that of silicon. Since silicon has the same type of structure as germanium (a diamond lattice) and is affected by chemical impurities in the same way as germanium, one might expect similar effects due to bombardment. This, however, is not so. Both *n*-type and *p*-type silicon increase in resistance (see Fig. 7). However, the increase in

† This effect of decay was found early in experiments with 20 MeV α-particles[32] and has recently also been observed after irradiation with polonium α-particles.[33]

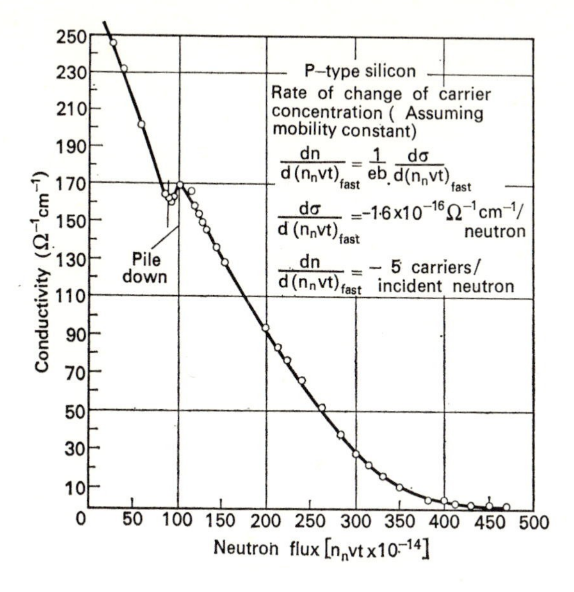

FIG. 7. Conductivity of p-type silicon as a function of fast neutron flux. Note effect of healing.

the p-type material is far greater than that in the n-type material[36] (see Tables 5 and 6). The explanation of this fact becomes clear when we consider heat treatment experiments and optical measurements carried out at Purdue,[37] which indicate that there are, in bombarded silicon, some deep-lying levels between 0·7 and 0·9 eV in the middle of the forbidden gap. If donors and acceptors in equal numbers are produced in a position in the middle of the band, where thermal ionization is rather inefficient, it can easily be seen that if one starts with an n-type conductor it loses all its conduction electrons to the acceptors. On the other hand, if one starts with a p-type conductor, then the electrons are lost by filling the holes in the full band. In both instances, resistivity increases.

The experiments indicate a much more rapid increase in resistivity in p-type, than in n-type, silicon. This indicates that the

TABLE 5. RESISTANCE CHANGES IN *p*-TYPE SILICON EXPOSED TO FAST NEUTRONS

Time (second)	0	10,000	20,000	30,000	40,000	52,000	62,000	82,000	100,000	158,000
Resistance (ohm)	29·8	49·9	59·2	78·6	120	190	463	3,518	16,090	68,000

TABLE 6. RESISTANCE CHANGES IN *n*-TYPE SILICON EXPOSED TO FAST NEUTRONS

Time (second)	0	5000	10,000	22,000	50,000	103,000	150,000	200,000	300,000	400,000
Resistance (ohm)	23·4	24·3	25·0	27·5	30·7	32·4	36·6	42·9	63·0	114·5

Fermi level is shifting towards a limiting position, determined by the imperfections alone, which lies in the upper half of the forbidden band.

The highest resistivity observed so far in the bombardment of silicon is only 10,000 ohm cm, which is still smaller by a factor of almost 30 than the expected intrinsic resistivity. X-ray studies[38] of the irradiated silicon do not show any new compounds formed during bombardment, except on the surface. The powder patterns of two irradiated samples have yielded results for lattice spacings, $5 \cdot 4185 \pm 0 \cdot 0005$ Å and $5 \cdot 4187 \pm 0 \cdot 0005$ Å, which are in agreement with the accepted values. Irradiation of silicon with deuterons produces the same effect as neutron irradiation: *p*-type material increases more rapidly in resistivity than *n*-type material. Heat treatment *in vacuo* at 450°C restores the original condition.

Electron Bombardment of Germanium

We have pointed out that primarily transient effects are to be expected from irradiation with electrons and γ-rays (see p. 189). However, M. M. Mills (North American Aircraft) pointed out to us that permanent effects may be obtained with electrons of high energy. Our experiments† with electrons may have given transient effects only because either insufficient energy was available to produce displacements or the currents used were so low that not enough displacements were produced to make the effect observable. If one estimates the order of magnitude of the recoil energy necessary to displace an atom completely in the lattice as 25 eV, then the 0·3 MeV electrons employed in our early work were below the threshold energy.

An estimate of the cross-section for this process, and of the time required to produce an observable effect, can be made by using Mott's[39, 40] theory. Let an electron (rest mass m) with energy W_p and a given momentum be incident on an atom of mass M_A. Then θ, the angle between the direction of incidence and the direction of recoil, is related to the recoil energy W_r by

† Experiments in collaboration with E. Klontz (Purdue).

$$\cos\theta = \frac{1 - \dfrac{W_r}{W_p}\left[1 + \dfrac{(M_A - m)\,c^2}{W_p - 2mc^2}\right]}{\left[1 - \dfrac{W_r}{W_p}\left(1 + \dfrac{W_p - W_r}{W_p + 2mc^2}\right)\right]^{\frac{1}{2}}} \tag{18}$$

With $W_r \ll W_p$ and $m \ll M_A$, this becomes approximately

$$\cos\theta = 1 - \frac{M_A c^2}{W_p + 2mc^2}\left(\frac{W_r}{W_p}\right) \tag{19}$$

For a head-on collision $\theta = \pi$ and approximately

$$W_r = \frac{2W_p m}{M_A}\left(\frac{W_p}{mc^2} + 2\right) \tag{20}$$

If W_r is 25 eV, then the threshold energy is 0·54 MeV. Using the expressions obtained in Mott's theory for the integrated cross-section

$$\begin{aligned}\sigma_\theta^{180} = \pi Z^2 \left(\frac{e^2}{mc^2}\right)^2 \left(\frac{1-\beta^2}{\beta^4}\right) \Bigg\{ &\cot^2\frac{\theta}{2} + 2\beta^2 \log\sin\frac{\theta}{2} \\ &+ \frac{2\pi\beta Z}{137}\left(\sin\frac{\theta}{2} + \operatorname{cosec}\frac{\theta}{2} - 2\right)\Bigg\} \\ &+ \text{higher order terms in } Z/137 \end{aligned} \tag{21}$$

one can estimate the currents and irradiation times necessary to produce observable effects. $\beta = v/c$. For 2 MeV electrons and for a current density of 1 μamp/cm^2 the time to produce an observable effect has been calculated as about 10 minutes.

Principle of experiment. In high resistivity *n*-type germanium the number of electrons to be "neutralized" is small, and if *n*-type germanium is converted to *p*-type, it should be possible to observe such an effect in a reasonable time. Qualitatively this can be detected simply by observing the change in rectification characteristic.

By bombarding partially shielded *n*-type germanium with electrons one can produce regions of *p*-type germanium bounded by n-type germanium (shielded portion). By observing the rectification characteristic, it is easily possible to distinguish between the two types of germanium as long as the thickness of the converted layer is large compared with the thickness of the barrier. This condition will be fulfilled for the energies in which we are interested, since even 0·3 MeV electrons, which produce no permanent effect, penetrate a germanium sample about 0·02 cm thick.

Experimental procedure. A given sample was irradiated at constant energy and current for varying intervals of time. A lead shield containing an aperture was placed over the sample so that only a small part was irradiated at one time. Several bombardments with varying time intervals and energies allow the threshold point to be determined approximately. This procedure also allows one to determine whether any effect observed was due to heating alone. The sample was finally probed and the shape of the rectifier characteristic for different parts of the sample observed on an oscilloscope. In order that each sample may be bombarded more than once, the samples chosen must be as uniform as possible.

The equipment and experimental arrangement are shown in Fig. 8. A rather large amount of power had to be dissipated in a very small volume of crystal. Therefore the sample holder was made from a heavy bar of copper, and the copper was immersed in liquid air during the bombardment. The bar was channelled and fitted with a spring and piston arrangement whereby the sample could be kept in thermal contact with the block and moved under the hole in the lead shield. In this manner, the region bombarded could be accurately controlled and after bombardment it was easy to probe a definite region. Figure 8a shows this arrangement. The irradiation hole was 3 mm square. Each sample had a copper lead soldered across one end, for contact in the probing and bombarding. In some bombardments the temperature was recorded. The rate of temperature rise was between 0·07 C°/sec (for 0·5 MeV electrons) and 0·2 C°/sec (1 MeV) so that the temperature was near 0°C at the end of a 30 minute bombardment.

During a single bombardment a heavy coating of frost encased the whole sample holder, and the sample adjusting arrangement froze solid. This necessitated allowing the temperature of the copper bar to come up above the freezing point between bombardments so that proper adjustments could be made.

Results. Six samples were subjected to bombardment in the course of these experiments. The results are shown in Table 7. We conclude that the threshold for the effect occurs between 0·5 and 0·7 MeV. Further bombardments will be necessary to establish its exact value. To determine whether an effect would occur in the time calculated, a sample was irradiated for 42 seconds at 2 MeV. The effect was still detectable. A *p*-type sample should exhibit a resistance decreasing as bombardment proceeds, just as with neutron bombardment. This has been found to be so.

These experiments show that it is possible to observe permanent effects with electron irradiation of higher energy. These preliminary experiments also indicate that the threshold energy necessary to remove a germanium atom from the lattice is about 30 eV, of the order of magnitude of the estimates found in the literature. Immediately after bombardment with 1 MeV and 0·7 MeV electrons for 15 minutes there is complete conversion to *p*-type. This effect "anneals" partially even at room temperature, by the time the characteristic is measured. Hence the final remark in Table 7 "poor" *p*-type characteristic. That ambient heating above room temperature cannot be responsible for the effect is clear, since we have subjected samples to electron bombardment from the electron gun of our electron diffraction equipment with a wattage far exceeding the one used with deuterons or in electron bombardment without producing any change in conductivity.

The fact that material bombardment at 0·5 MeV for 30 minutes showed no effect also indicates that the bombardment effects are not due to simple heat treatment. Many more experiments of a quantitative nature will be necessary to decide whether ionization processes or "spot heating" may not also account for such an effect. They may also lead to a lower threshold for the permanent change of *n*-type germanium to *p*-type germanium. The very small

TABLE 7. RESULTS OF CONTINUOUS BOMBARDMENT OF *p*- AND *n*-TYPE GERMANIUM BY HIGH ENERGY ELECTRONS

Sample No.	Resistivity ϱ ohm cm	Bombardment		Beam		Energy to sample joule	Results of bombardment	
		No.	Time sec	Energy MeV	Current* μamp		ϱ of incident face ohm cm	Changes in type
n–1	11·6	1 2	200 2,000	2·0 2·0	10 10	400 4,000	~ 26 32	Change to *p*-type
n–2	4·75	1 2 3	300 1,200 2,000	1·5 1·0 0·7	10 10 9	450 1,200 1,260	44 13 5	Fair to poor *p*-type
		4	2,400	0·5	6	720	Same	No change
n–3	16·3	1	42	2·0	10	84	40	Poor *p*
n–4	1·4	1	900	1·0	1·7†	1,530	2·1	Poor *p*
n–5	5·0	1 2	900 1,800	0·7 0·5	1·0† 1·0†	630 900	— —	Poor *p* No change
p–1	6·2	1	350	2·0	10	700	4·0	ϱ decreased

* Current to sample ~ 0·1 total current. † Actual current to sample.

number of impurity centres which has to be produced to give a measurable electrical effect shows how much more sensitive electrical measurements are for the detection of changes of structure in semiconductors than they are in metals and other materials.

Transient Processes

The possibility of producing conduction in a semiconductor by electron irradiation or ionizing radiation has been discussed by various authors.[41–44] In connection with our experiments on nucleon bombardment, we have irradiated germanium samples with electrons of about 0·3 MeV energy from our Van de Graaff generator. The germanium samples, 0·02 cm thick, were "transparent" to this radiation; thus volume effects were observed.[7]

An *n*-type, 18 ohm cm sample of germanium was enclosed in an evacuated chamber attached to the end of the discharge column of the electrostatic generator. The sample was irradiated with 340 keV electrons in 1 μsec pulses produced by a thyratron circuit. The pulse frequency was 200 per second and the current 300 μamp per pulse. A 2200 ohm resistance and a DC supply were connected in series with the sample. Variations in time of the current through the sample and the voltage across the sample were observed by means of an oscilloscope, the sweep of which was synchronized with the pulsing apparatus. From these variations, calculation of the resistance as a function of time showed a drop from 1460 to 400 ohm during each pulse. This change was not permanent and recovery to the initial resistance was complete, but not instantaneous, requiring some 10 μsec for the process (see Fig. 9).

p–n Barriers as counters. In some of the experiments on the production of *p–n* boundaries in germanium by irradiation with deuterons and α-particles, it was found that barrier characteristics change after bombardment ceases. We have therefore investigated the possibility that this might be due to radioactivity induced in the material, and whether such a *p–n* boundary might act as a counter for ionizing radiation.[45] This has been found to be so.

The author has investigated both natural *p–n* barriers produced in the melt, and artificial *p–n* barriers produced by deuteron bombardment. The natural *p–n* boundaries had an extension of $1 \times 0{\cdot}3 \times 0{\cdot}1$ cm, whereas the artificial barrier had an extension of $0{\cdot}5 \times 0{\cdot}2 \times 0{\cdot}02$ cm. The behaviour of both natural and artificial barriers is essentially the same, except for the differences introduced by the physical dimensions, namely, the transparency

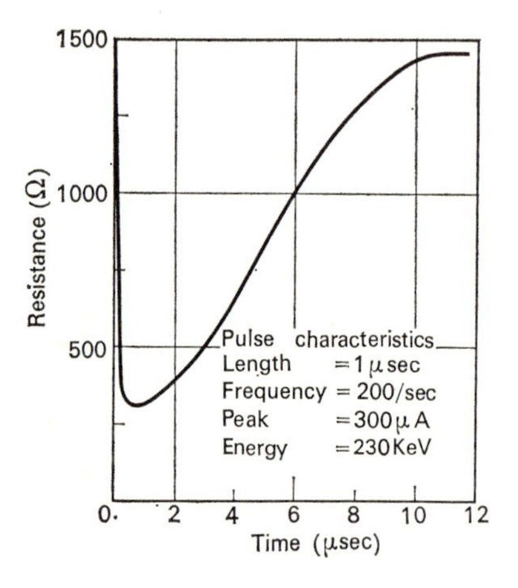

FIG. 9. Resistance changes in *n*-type germanium (ϱ = 18 ohm-cm) under pulsed electron bombardment.

to β- and γ-radiation. The effective counting area is of the order of 10^{-4} cm^2. The time of rise is less than $0{\cdot}05$ μsec. The maximum pulse height is about 2 mV.

Since the *p–n* boundary is highly light-sensitive, it is essential that the boundary be kept in a thermostat which is completely light-tight. The barrier was investigated at various temperatures from room temperature down to liquid nitrogen temperature. At room temperature, the temperature was controlled to about $\pm 0{\cdot}05$ C°. The load resistance was 10^5 ohm (with a total voltage drop of 100 V) and the bias across the barrier was 10 V, giving a bias current of about 1 ma. At liquid air temperature the barrier resistance is of the order of 100 megohm. The barrier is very sensitive to room temperature radiation, and therefore has to be carefully shielded. It is sensitive to α-, β- and γ-radiation. After exposure of the barrier to more than 10^7 α-particles, the maximum of the pulse height distribution shifts slowly to lower values. The barrier is stable to small temperature changes, but the maximum of pulse height distribution is very sensitive to slight temperature shifts. The pulse height distribution is typical of crystal counters of small cross-section. It shows a very sharp maximum with α-particles, but the upper limit is smeared out with β- and γ-particles.

The *p–n* barrier is a most interesting object for the study of the mechanism of crystal counters in general. A large number of parameters is variable, such as resistance, dimensions, etc., which are not as easily variable with ionic crystal counters. The number of carriers in the barrier can be varied at will either by adjusting the temperature of the sample, or by irradiating it with light. In addition, we hope by a more careful and quantitative investigation of the counting mechanism to get more insight into the model of the *p-n* barrier and the germanium semiconductor when exposed to ionizing radiation in general.

Summary

1. Transmutations produced in germanium by slow neutrons lead to excess *p*-type production, and the cross-sections determined with separated germanium isotopes indicate that the number of excess *p*-type impurities to be expected is 2% of the total slow flux. It has been found that, with slow neutrons, approximately three times as many *p*-type impurities (gallium) are produced as *n*-type impurities (arsenic).

Comparing the results for integrated flux with Hall effect measurements after irradiation and heat treatment, it can be shown that each impurity centre releases one carrier, and that *n*-type and *p*-type impurity centres are equally effective. By irradiating for limited periods and gradually increasing the amount of irradiation, it is possible to produce material which has the properties of intrinsic semiconductors because of a balance between the two types of impurities.

2. *n*-Type germanium is converted into *p*-type germanium, and *p*-type germanium conductivity increases, upon irradiation with α-particles, deuterons, neutrons, and fast electrons. By using well defined beams of charged particles it is possible to produce a *p–n* boundary in pre-determined locations.

The resistance of *n*-type material reaches a maximum when the conduction electrons are "neutralized" and the resistivity then decreases again when the material is converted to *p*-type. The number of displacements can be calculated both for charged particles and for neutrons, but the actual number of carriers found is much smaller than the number of displacements. This is due to a healing effect and also because the acceptor and donor levels produced by lattice defects partially cancel.

Analysis of the neutron irradiation curves shows that in *n*-type germanium fast neutrons remove 3 carriers per incident neutron, while analysis of irradiation experiments with *p*-type material show that, depending on the temperature, about 0·7 carrier is added per incident neutron. This indicates that defects produced by bombardment act as donors and acceptors with activation energies larger and distributed over a wider range of energy than those observed with chemical impurities.

It can be shown that the number of transmutations produced with deuterons cannot account for the large conductivity effects found by irradiating with 10 MeV deuterons. Heat treatment at 450°C restores the original condition of the semiconductor bombarded as described in this section.

3. Experiments with electrons of energies from 0·7 MeV up produce the same kind of permanent effects in germanium as heavy particles: changes from *n*- to *p*-type. From the lower limit of energy necessary to produce this effect one can calculate a minimum energy necessary to remove a germanium atom completely from its lattice site. This is found to be about 30 eV.

Experiments with 0·5 MeV electrons produce no such effects and indicate that ambient heating cannot be responsible for the effects observed at 0·7 MeV. This is also checked by experiments with prolonged irradiation with electrons of much lower energy (50 keV).

4. Using electrons of energy lower than 0·7 MeV and proper oscillographic equipment, it is possible to show that such irradiation of germanium produces conductivity pulses and a decrease in resistance up to a factor of about 10, but these conductivity pulses do not last longer than about 10 μsec.

Transient effects are observed also when both natural and artificially prepared *p–n* boundaries in germanium are exposed to ionizing radiation. It is possible to produce a crystal counter which responds at room temperature to α-, β- and γ-radiation. The time of rise is less than 0·05 μsec, and the maximum pulse height is about 2 mV with a bias of not more than 10 V.

Acknowledgements

We are indebted to E. Bleuler and D. Tendam for help and advice in the cyclotron bombardment experiments. We are indebted to G. E. Evans, H. Y. Fan, H. Fröhlich, and H. M. James for many valuable discussions on solid state problems involved in the irradiation experiments.

The work at Purdue University was sponsored by a Signal Corps contract. The work at the Oak Ridge National Laboratory was carried on by the author in collaboration with the solid state group as consultant to the Physics Division.

We wish to express our appreciation to F. F. Rieke and J. W. MacKay of the Purdue electron accelerator staff for their assistance in making the first experiments on electron bombardment; to Dr. B. Waldman and members of the staff of the Notre Dame electrostatic generator whose co-operation and assistance made possible the successful outcome of these experiments, and to the Van de Graaff group at the Argonne National Laboratory, O. C.

Simpson, J. R. Gilbreth and S. M. Black, where the experiments are being continued at the present time.

References

1. Lark-Horovitz, K. *Nat. Def. Res. Comm. Rep.* No. 14–585, pp. 7–57, covering March 1942 to Nov. 1945.
2. Bardeen, J. and Pearson, G. L. *Phys. Rev.* **75**, 865 (1949).
3. Scaff, J. H., Theuerer, H. C. and Schumacher, E. E. *Trans. Amer. Inst. min. (metall.) Engrs* **185**, 383 (1949).
4. Bardeen, J. and Brattain, W. *Phys. Rev.* **75**, 1216 (1949).
5. Bray, R. *Purdue Prog. Rep.* Feb. 1946 to Jan. 1949, p. 70.
6. Bray, R. *Ibid.* Sept. 1949 to Nov. 1949, p. 62.
7. Klontz, E. and Lark-Horovitz, K. *Progr. Rep.*, Contract No. W36-039-32020 (Signal Corps) Aug. to Oct. 1948, p. 41.
8. Davis, R. E. and Lark-Horovitz, K. *Progr. Rep.*, Contract No. W36-039-32020 (Signal Corps) Nov. 1947 to Jan. 1948.
9. Lark-Horovitz. K., Bleuler, E., Davis, R. E. and Tendam, D. L. *Phys. Rev.* **73**, 1256 (1948).
10. Davis, R. E., Johnson, W. E., Lark-Horovitz, K. and Siegel, S. *Ibid.* **74**, 1255 (1948).
11. Davis, R. E., Johnson, W. E., Lark-Horovitz, K. and Siegel, S. *A.E.C.D. Report* No. 2054, 1948.
12. Johnson, W. E. and Lark-Horovitz, K. *Phys. Rev.* **75**, 442 (1949).
13. James, H. M. Private communication.
14. Klontz, E. and Lark-Horovitz, K. *Progr. Rep.* Dec. 1949 to Feb. 1950, p. 52.
15. Pomerance, H. L. Private communication.
16. Cleland, J., Lark-Horovitz, K. and Pigg, J. C. *Phys. Rev.* **78**, 814 (1950).
17. Cleland, J., Crawford, J. H., Lark-Horovitz, K. and Pigg, J. C. *O.R.N.L. Quart. Rep.* June 1950.
18. Weisskopf, V. F. *Los Alamos Lectures*, 1946.
19. Bleuler, E. and Tendam, D. J. *Progr. Rep.* Nov. 1947 to Jan. 1948.
20. Bohr, N. *K. danske vidensk, Selsk. Skr.* **8**, 8 (1948).
21. Allen, E. O. and Seitz, F. Irradiation Chemistry and Physics, *O.R.N.L. Rep.*, 1946.
22. Seitz, F. *Discuss. Faraday Soc.* **5**, 271 (1949).
23. Evans, G. E. Private communication.
24. Dessauer, F. *Z. Phys.* **12**, 38 (1923).
25. Brooks, H. and Lawson, A. L. Private communication.
26. James, H. M. Private communication.
27. Cleland, J. and Lark-Horovitz, K. *Progr. Rep.* June 1949 to Aug. 1949, p. 106.
28. Crawford, J. H. and Lark-Horovitz, K. *Phys. Rev.* **78**, 815 (1950).
29. Crawford, J. H. and Lark-Horovitz, K. *Ibid.* In press.
30. Johnson, V. A. and Lark-Horovitz, K. *Ibid.* **79**, 176 (1950).

31. Shockley, W., Pearson, G. L. and Haynes, J. R. *Ibid.* **78,** 295 (1950).
32. Davis, R. E. and Lark-Horovitz, K. *Progr. Rep.* Nov. 1947 to Jan. 1948, p. 43.
33. Brattain, N. H. and Pearson, G. L. *Bull. Amer. phys. Soc.* **25,** 15 (1950).
Brattain, N. H. and Pearson, G. L. *Phys. Rev.* **78,** 646 (1950).
34. Crawford, J. H. *O.R.N.L. Quart. Rep.* In preparation.
35. Cleland, J., Crawford, J. H., Lark-Horovitz, K. and Pigg, J. C. *O.R.N.L. Rep.* March 1950.
36. Johnson, W. E. and Lark-Horovitz, K. *N.E.P.A. Radiation Damage Symposium*, Dec. 1948, Rep. No. 1178–IER-23.
37. Lark-Horovitz, K., Becker, M., Davis, R. E. and Fan, H. Y. *Phys. Rev.* **78,** 334 (1950).
38. Geib, I. G. Private communication.
39. Mott, N. F. *Proc. roy. Soc., Lond.* **A124,** 429 (1929).
40. Barber, A. and Champion, F. C. *Ibid.* **168,** 159 (1938).
41. Rittner, E. S. *Phys. Rev.* **73,** 1212 (1948).
42. McKay, K. G. *Ibid.* **74,** 1606 (1948).
43. McKay, K. G. *Ibid.* **76,** 1537 (1949).
44. Ansbacher, A. and Ehrenberg, W. *Nature, Lond.* **164,** 144 (1949).
45. Orman, C., Fan, H. Y., Goldsmith, G. J. and Lark-Horovitz, K. *Bull. Amer. phys. Soc.* **25,** 15 (1950).
Orman, C., Fan, H. Y., Goldsmith, G. J. and Lark-Horovitz, K. *Phys. Rev.* **78,** 646 (1950).

APPENDIX I

Effect of Bombardment upon a Classical Semiconductor in Thermal Equilibrium†

The behaviour of germanium under neutron bombardment is quite similar to the behaviour of lead sulphide under the action of oxygen. Just as in bombarded germanium, we observe in lead sulphide transitions from *n*-type to *p*-type and the rise of resistivity to a maximum, followed by a decrease when the material has become *p*-type by the introduction of sufficient oxygen. As a consequence, the following considerations of the behaviour, under thermal equilibrium, of the conductivity, Hall effect, and thermoelectric power as a function of bombardment are of interest beyond the specific case of neutron bombardment of a semiconductor.

† By V. A. Johnson and K. Lark-Horovitz.

If we assume that classical statistics† can be used and that the semiconductor is in thermal equilibrium, then one can write for the conductivity

$$\sigma = n_e e b_e + n_h e b_h$$

$$= e b_h (n_e \Lambda + n_h) = e b_h \left(\frac{\Lambda K}{n_h} + n_h \right) \tag{22}$$

where Λ is again the ratio of electron to hole mobility, and $n_h n_e = K$ the equilibrium constant for a given fixed temperature. Likewise, one can write for the Hall coefficient

$$\mathscr{R} = \frac{3\pi}{8e} \cdot \frac{n_h b_h^2 - n_e b_e^2}{(n_h b_h + n_e b_e)^2} = \frac{3\pi}{8e} \cdot \frac{(n_h - \Lambda^2 n_e)}{(n_h + \Lambda n_e)^2} = \frac{3\pi\, n_h (n_h^2 - \Lambda^2 K)}{8e\, (n_h^2 + \Lambda K)^2} \tag{23}$$

and for the thermo-electric power

$$\frac{dE}{dT} = -\frac{k}{e} \left[2(n_e \Lambda - n_h) - n_e \Lambda \log \frac{n_e h^3}{2(2\pi m_e^* kT)^{3/2}} + n_h \log \frac{n_e h^3}{2(2\pi m_h^* kT)^{3/2}} \right] (n_e \Lambda + n_h)^{-1} \tag{24}$$

The conductivity reaches a minimum value when

$$n_h = \Lambda n_e = (\Lambda K)^{\frac{1}{2}} \tag{25}$$

The minimum conductivity is given by

$$\sigma_{\min} = 2 e b_h (\Lambda K)^{\frac{1}{2}} \tag{26}$$

For comparison, note that the intrinsic conductivity at the same temperature is given by

$$\sigma_i = (1 + \Lambda) e b_h K^{\frac{1}{2}} \tag{27}$$

† At 300°K the Fermi level is below the conduction band (or above the full band) by $3kT$ or more for all carrier concentrations below 10^{18} per cm^3. This condition ensures the applicability of classical statistics to the type of sample considered here.

and thus†
$$\frac{\sigma_{min}}{\sigma_i} = \frac{2\Lambda^{\frac{1}{2}}}{1 + \Lambda} \tag{28}$$

Thus $\sigma_{min} = \sigma_i$ if $\Lambda = 1$, $\sigma_{min} < \sigma_i$ if $\Lambda > 1$, and $\sigma_{min} > \sigma_i$ if $\Lambda < 1$.

Now consider the value of the Hall coefficient at the time that $\sigma = \sigma_{min}$. Upon putting $n_h = (\Lambda K)^{\frac{1}{2}}$ and $n_e = (K/\Lambda)^{\frac{1}{2}}$, one obtains

$$\mathscr{R}' = \frac{3\pi}{32e} \cdot \frac{1 - \Lambda}{(\Lambda K)^{\frac{1}{2}}} \tag{29}$$

Thus $\mathscr{R}'$ is negative, zero, or positive depending upon whether Λ is greater than, equal to, or less than unity. In germanium the mobility ratio Λ is > 1 and thus one observes that, when the conductivity has reached its minimum, the Hall effect is still negative in sign even though the number of holes is greater than the number of electrons. The Hall coefficient passes through zero to become positive when $n_h = \Lambda K^{\frac{1}{2}}$ and $n_e = K^{\frac{1}{2}}\Lambda$.

Likewise, if one examines the thermo-electric power when $\sigma = \sigma_{min}$, it is found that

$$(dE/dT)' = (-\,k/2e) \log (\Lambda m_e^{3/2}/m_h^{3/2}) \tag{30}$$

This reduces to

$$(dE/dT)' = (-\,k/2e) \log (m_h/m_e) \tag{31}$$

if one considers the mobility ratio[46] Λ to equal $(m_h/m_e)^{5/2}$. Thus it is found that (dE/dT) is negative, zero, or positive depending on whether m_h is greater than, equal to, or less than m_e; the latter conditions correspond to $\Lambda > 1$, $\Lambda = 0$, and $\Lambda < 1$, respectively.

If it is assumed that the semiconductor remains in thermal equilibrium, a detailed study can be made of Hall effect and

† This relation is true if b_h is the same for σ_{min} and σ_i. The effect of bombardment in producing additional scattering centres tends to reduce b_h for σ_{min} and thus to make the ratio σ_{min}/σ_i less than indicated. The effect depends on n, the initial concentration of free electrons from impurities. There is no appreciable decrease of b_h for $n \leqslant 10^{14}/cm^3$, about 1 per cent decrease for $n = 10^{15}/cm^3$ and 10 per cent decrease for $n = 10^{16}/cm^3$.

conductivity behaviours as functions of bombardment. Calculations of this type have been carried out in simplified form, in connection with the discussion of lead sulphide, by Putley,[47] and by Fry[48] at the Atomic Energy Research Establishment. Detailed calculations of the Fermi level in connection with the lead sulphide problem have been made by Müser.[49] It is assumed that one starts with an *n*-type germanium sample, that the bombardment produces a defect density proportional to the integrated flux, and that electrons from the conduction band fall into the acceptor levels associated with the defects, causing an eventual decrease in free electrons to practically zero and a subsequent production of holes in the full band, i.e. a conversion of the sample from *n*- to *p*-type.

The thermal equilibrium condition[50] is that

$$n_h n_e = K = 3{\cdot}61 \times 10^{26}\ \text{cm}^{-6} \text{ for } 300°\text{K} \tag{32}$$

Before the bombardment starts, $n_e = N + n$ and $n_h = n$, where N is the initial electron density due to impurities and n is the corresponding density of intrinsic electrons or holes. Let N' be the defect density due to bombardment and subsequent transmutation. Then, as long as $N' \leqslant N$, one can write

$$n_e = N - N' + n,\ n_h = n \tag{33}$$

Note that n varies with N and N' so that $n_e n_h = K$ at all times. When $N' \geqslant N$, one takes

$$n_e = n,\ n_h = n + N' - N \tag{34}$$

During the first part of the bombardment, as long as $N' < N$, the Hall coefficient is given by

$$\mathscr{R} = \frac{r_c}{e}\frac{[(1 - \Lambda^2)(K + \mathscr{N}_1^{\,2}/4)^{\frac{1}{2}} - (1 + \Lambda^2)\mathscr{N}_1/2]}{[(1 + \Lambda)(K + \mathscr{N}_1^{\,2}/4)^{\frac{1}{2}} + (\Lambda - 1)\mathscr{N}_1/2]^2} \tag{35}$$

where $N_1 = N - N'$ and r_c is a numerical factor of the order of unity, for which the exact value depends upon the nature of the scattering encountered by the carriers.

When $N' = N$, one finds that

$$\mathscr{R} = (-r_c/eK^{\frac{1}{2}})(\Lambda - 1)/(\Lambda + 1) \quad (36)$$

Thus the Hall effect is still negative if $\Lambda > 1$.

When $N' > N$, the Hall coefficient becomes

$$\mathscr{R} = \frac{r_c}{e} \frac{[(1 - \Lambda^2)(K + \mathscr{N}_2^2/4)^{\frac{1}{2}} + (1 + \Lambda^2)\mathscr{N}_2/2]}{[(1 + \Lambda)(K + \mathscr{N}_2^2/4)^{\frac{1}{2}} - (\Lambda - 1)\mathscr{N}_2/2]^2} \quad (37)$$

where $N_2 = N' - N$.

Corresponding expressions for the conductivity are

$$\sigma = eb_e[(1 + 1/\Lambda)(K + \mathscr{N}_1^2/4)^{\frac{1}{2}} + (1 - 1/\Lambda)\mathscr{N}_1/2] \quad (38)$$

and

$$\sigma = eb_e[(1 + 1/\Lambda)(K + \mathscr{N}_2^2/4)^{\frac{1}{2}} - (1 - 1/\Lambda)\mathscr{N}_2/2] \quad (39)$$

Figure 10 shows the variation in $\mathscr{R}$ and σ with N' as calculated for a germanium sample with $N = 10^{16}/\text{cm}^3$. In obtaining these results, r_c was taken as one, Λ as $1 \cdot 50$, and b_c as 2800 cm²/volt sec.

The expressions for $\mathscr{R}$ are differentiated to find the locations of the minimum and maximum. These positions depend on Λ and K. The results are:

$$\mathscr{R} = \mathscr{R}_{\text{min}}^{-} \text{ for } \mathscr{N}_1 = 1 \cdot 245 \times K^{\frac{1}{2}} = 2 \cdot 36 \times 10^{13}/\text{cm}^3 \text{ and}$$

$$\mathscr{R}_{\text{min}}^{-} = -\frac{0 \cdot 330r}{eK^{\frac{1}{2}}} \approx -108{,}000 \text{ cm}^3/\text{coulomb}.$$

Also

$$\mathscr{R} = \mathscr{R}_{\text{max}}^{+} \text{ for } \mathscr{N}_2 = 3 \cdot 005\, K^{\frac{1}{2}} = 5 \cdot 71 \times 10^{13}/\text{cm}^3 \text{ and}$$

$$\mathscr{R}_{\text{max}}^{+} = \frac{0 \cdot 1858r}{eK^{\frac{1}{2}}} \approx 61{,}000 \text{ cm}^3/\text{coulomb}.$$

Thus, with the progression of the bombardment, one observes

the negative Hall coefficient increasing in magnitude to a maximum while the conductivity decreases, the negative Hall constant starts to decrease in magnitude while the conductivity continues to decrease for a while longer, then the conductivity reaches its minimum and starts to increase before the Hall coefficient reaches zero, and the conductivity continues to increase as the Hall curve passes through a positive maximum and then decreases.

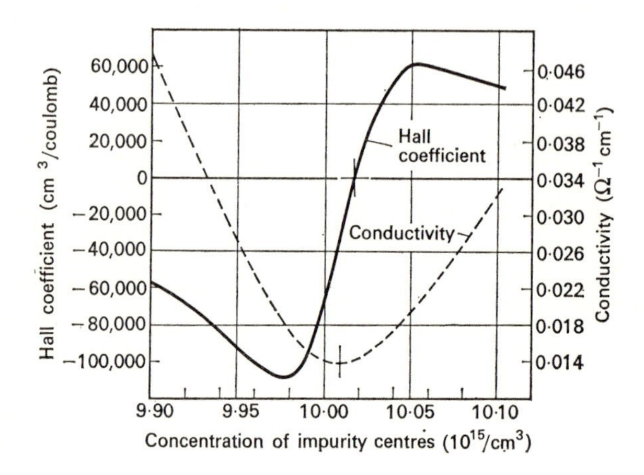

FIG. 10. Hall coefficient and conductivity of *n*-type germanium in temperature equilibrium as a function of the concentration of impurity centres introduced by neutron irradiation.

REFERENCES

46. LARK-HOROVITZ, K. *Contractor's Final Rep. N.D.R.C.* No. 14–585, November 1945, p. 49.
47. PUTLEY, E. H. *T.R.E. Memo.* No. 235 (1950).
48. FRY, T. M. Private communication.
49. MÜSER, H. *Z. Naturf.* **5a**, 18 (1950).
50. JOHNSON, V. A. and FAN, H. Y. *Phys. Rev.* **79**, 899 (1950).
51. SCANLON, W. and LARK-HOROVITZ, K. *Ibid.* **73**, 1256 (1948).

APPENDIX II

FIRMI LEVELS IN BOMBARDED SEMICONDUCTORS†

In discussing the behaviour of the Fermi level in bombarded semiconductors it will be assumed here that the donor states introduced by bombardment all have the same activation energy, and likewise that the acceptor states all have the same energy.

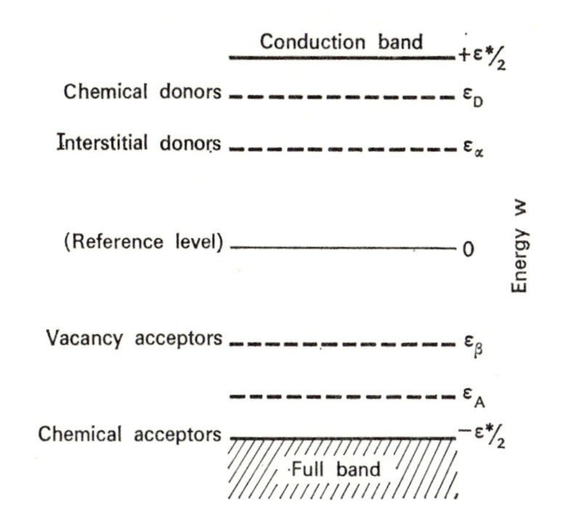

FIG. 11. Band diagram for a bombarded semi-conductor.

The notation to be used is illustrated in Fig. 11. The zero of energy is taken in the middle of the forbidden band, which has width ϵ^*; thus the adjacent edge of the conduction band has energy $+ \epsilon^*/2$, and the adjacent edge of the full band has energy $- \epsilon^*/2$. A constant density N_+ of chemically introduced donor levels with energy ϵ_D will be assumed to be present, and a constant density N_- of chemically introduced acceptor levels with energy ϵ_A. The bombardment will introduce donor levels with energy ϵ_α, and acceptor levels with energy ϵ_β. The total concentration of donors and acceptors introduced by the bombardment will be N';

† By H. M. James and G. W. Lehman.

of these a fraction f_α will be donors, and a fraction $f_\beta = 1 - f_\alpha$ will be acceptors.

If μ is the Fermi level, the probability that an n-type donor is ionized will be

$$P^+(w;\mu) = \frac{1}{1 + \exp[-(w-\mu)/kT]} \tag{40}$$

and the probability that a p-type acceptor will be ionized is

$$P^-(w;\mu) = \frac{1}{1 + \exp[(w-\mu)/kT]} \tag{41}$$

w being in each case the energy of the appropriate level. The concentration of electrons in the conduction band is given by

$$n_e(\mu) = \frac{4\pi(2m_e^*)^{3/2}}{h^3}\int_0^\infty \frac{w^{\frac{1}{2}}\,dw}{1 + \exp[(w + \epsilon^*/2 - \mu)/kT]} \tag{42}$$

and the concentration of holes in the filled band by

$$n_h(\mu) = \frac{4\pi(2m_h^*)^{3/2}}{h^3}\int_0^\infty \frac{w^{\frac{1}{2}}\,dw}{1 + \exp[(w + \epsilon^*/2 + \mu)/kT]} \tag{43}$$

where m_e^* and m_h^* are the effective masses of the electrons and holes respectively.

The position of the μ level is determined by the condition of electrical neutrality of the crystal

$$n_h(\mu) - n_e(\mu) + N_+ P^+(\epsilon_D;\mu) - N_- P^-(\epsilon_A;\mu) + N'[f_\alpha P^+(\epsilon_\alpha;\mu) - f_\beta P^-(\epsilon_\beta;\mu)] = 0 \tag{44}$$

Three special models will be considered here, all with $\epsilon^* = 0{\cdot}75$ eV, corresponding roughly to the band width in germanium and with $m_e^* = m_h^* = m$. The only temperature considered is 20°C.

Model I. It is assumed that the chemical impurities present are completely ionized, so that the condition on μ becomes

$$n_e(\mu) - n_h(\mu) + N_- - N_+ = N'[f_\alpha P^+(\epsilon_\alpha;\mu) - f_\beta P^-(\epsilon_\beta;\mu)] \tag{45}$$

In each case only one type of chemical impurity is assumed to be present.

The levels introduced by bombardment are taken as $\epsilon_\alpha = 0 \cdot 175$ eV, $\epsilon_\beta = -0 \cdot 175$ eV. The values of f_α considered are 0·4, 0·5 and 0·6, corresponding to excess of acceptors, equality in number of acceptors and donors, and excess of donors, respectively.

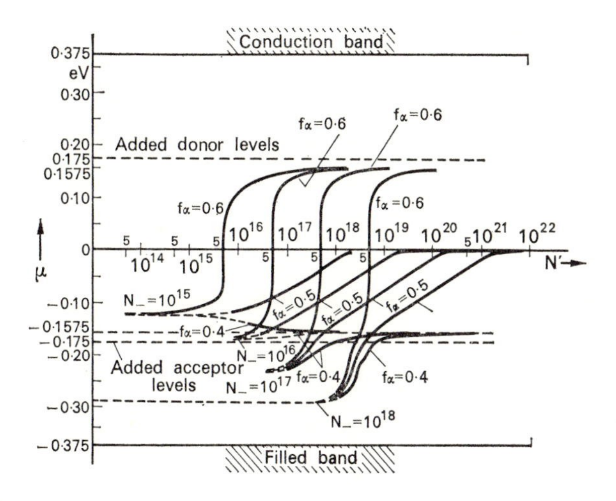

FIG. 12. Position of the Fermi level as a function of N' for a material that is p-type before bombardment.

Figure 12 shows a plot of μ against N' for material that is p-type before bombardment. Corresponding curves for n-type material are obtained on reflecting these in the reference level and replacing N_- by N_+, f_α by $1 - f_\alpha$.

If $f_\alpha = 0 \cdot 6$, three donor levels are introduced for each pair of acceptors, and the material changes to n-type during bombardment. The excess of donors over acceptors is $N'/5$. It will be observed that in each case the μ level passes through $w = 0$ when $N'/5$ is essentially equal to the initial excess of acceptors. The limiting position of the μ level for large N', $0\cdot01575$ eV, is determined entirely by the character of the levels added by bombardment, and corresponds to a probability of about 2/3 for ionization of the donors.

If $f_\alpha = f_\beta = 0\cdot5$ the μ level approaches the middle of the band, because of the symmetry of this model. Here the resistance of the material would increase monotonically during bombardment, whether it is initially *n*-type or *p*-type.

If $f_\alpha = 0\cdot4$, bombardment adds an excess of acceptors, and the material remains *p*-type. It is interesting to note that, nevertheless, the μ level will rise and the resistance will increase if the initial concentration of completely ionized acceptors is high. This happens because the donors fill most of the holes in the "full" band; for the most part the holes in the bombarded material are trapped in the acceptor levels, which are not easily ionized.

Model II. Here it is assumed that the chemically introduced acceptors have zero activation energy: $\epsilon_D = -\epsilon_A = 0\cdot375$ eV. (Note that this does not imply complete ionization.) Only one type of chemical impurity is assumed to be present in each case. It is assumed that the bombardment produces equal concentrations of donor and acceptor levels ($f_\alpha = f_\beta = 0\cdot5$) with energies $\epsilon_\alpha = -0\cdot225$ eV and $\epsilon_\beta = -0\cdot325$ eV, respectively.

Figure 13 shows plots of μ against $N'/2$, the number of added donor or acceptor levels, for various initial concentrations of chemical impurities. In all cases the μ level approaches the limit $w = -0\cdot275$ eV as N' increases.

The behaviour of this model is similar to that observed with germanium: initially *n*-type material is converted to *p*-type material, and initially *p*-type material shows decreasing resistance if the concentration of chemical acceptors is moderate. Very impure *p*-type material would, on the other hand, show increasing resistance as the bombardment proceeds. One would expect the resistance to approach the same limit in all cases in which the mobilities are the same.

Model III. This model is intended to simulate the behaviour of germanium containing impurities produced by bombardment-

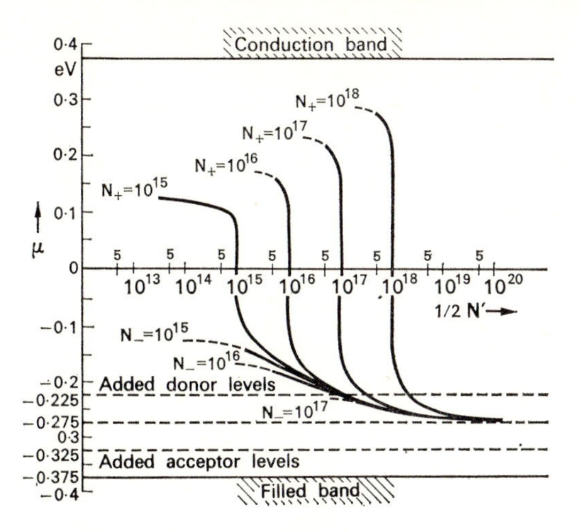

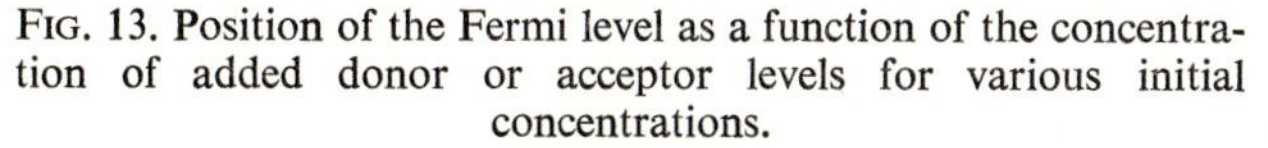
FIG. 13. Position of the Fermi level as a function of the concentration of added donor or acceptor levels for various initial concentrations.

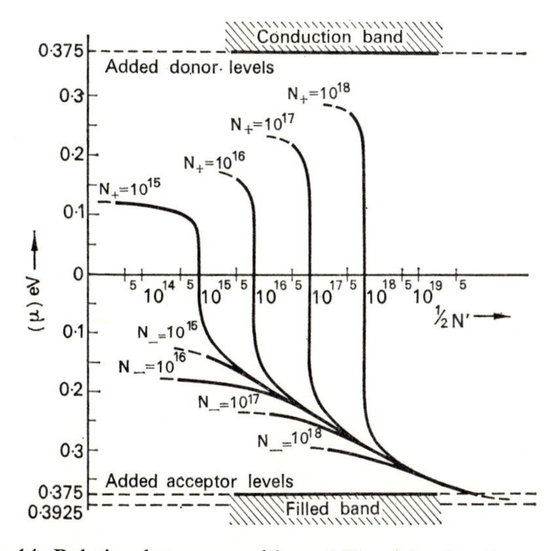

FIG. 14. Relation between position of Fermi level and excess of acceptors over donors induced by transmutation.

induced transmutations, when interstitials and vacancies have been removed by heat treatment.

The original impurities, either *n*- or *p*-type, are taken to have zero activation energy; so are the *n*- and *p*-type impurities introduced by bombardment in a ratio taken to be 1 to 3. The μ level is then determined by

$$n_e(\mu) - n_h(\mu) + [N_- + \tfrac{3}{4}N'] \, P^+(-\,0\cdot375;\mu) - [N_+ + \tfrac{1}{4}N'] \, P^-(+\,0\cdot375;\mu) = 0 \qquad (46)$$

Figure 14 shows a plot of μ against $N'/2$, the excess of acceptors over donors. Again it will be noted that *n*-type material is converted to *p*-type, and *p*-type material shows decreasing resistance. In all cases the limiting position of the μ level is at $-\,0\cdot3925$ eV below the edge of the filled band.

16

Localized Electronic States in Bombarded Semiconductors†[1]

ABSTRACT

Bombardment of germanium and silicon with fast α-particles, deuterons, neutrons or electrons produces striking changes in the electrical and optical properties of these materials. Bombardment of *n*-type germanium produces a decrease in conductivity (linear at the start), which finally passes through a minimum and begins to increase after the material has become *p*-type; bombarded *p*-type germanium shows a monotone increase in conductivity unless this is initially very high. This behavior can be explained by the assumption that the bombardment produces more acceptors than donators in the material. On the other hand, bombarded silicon, whether of *n*-type or *p*-type, shows monotonically increasing resistance; this can be attributed to introduction of both electron traps and hole traps by the bombardment.

It is the purpose of this paper to develop a more complete basis for the understanding of these and other properties of bombarded semiconductors. A discussion of the effects of substitutional impurities is followed by an analysis of the action of vacancies in polar crystals as acceptors and donators. It is then argued that lattice vacancies in germanium and silicon will act as acceptors, and interstitial atoms will act as donators, for more than one electron; in other words, interstitials will show several successively increasing electron ionization energies, and vacancies several successively increasing hole ionization energies. Analysis of observations on bombarded germanium indicates that the second electron ionization energy of germanium interstitials is of the order of the forbidden band width, 0·75 eV, and that the second hole ionization energy is low, of the order of 0·2 eV; it then appears that in bombarded germanium the ζ-level is determined by the equilibrium between singly and doubly ionized interstitials and vacancies. In the case of silicon, it appears that the second hole ionization energy is much larger (~ 0·75 eV) and that the second electron ionization energy is increased to about 1 eV; in bombarded silicon most of the interstitials and vacancies will be singly ionized.

† *Zeits. f. phys. Chemie* **198,** 107 (1951), with H. M. James.

I. Introduction

The importance of lattice imperfections for the electrical properties of semi-conductors was recognized some time ago.[2] These imperfections may be either substitutional impurities or lattice imperfections such as vacancies and interstitial atoms.

In germanium and silicon it has been found[3] that addition of elements from the third column of the periodic system (one valence electron less than in germanium and silicon) produces hole-conduction: *p*-type material. Addition of elements of the fifth column of the periodic system (one valence electron more than in germanium and silicon) produces conduction by electrons: *n*-type material. That there is one carrier released for each such impurity atom added is shown by analysis of the temperature dependence of resistivity[3], and by experiments with radioactive indicators[4] and impurity atoms produced *in situ* by transmutation with slow neutrons in the nuclear reactor.[5]

On quenching *n*-type germanium from about 800°C, *p*-type germanium is produced, but by annealing at about 450°C and slowly cooling the material one can reconvert it to *n*-type material.[6] These effects, due to heat treatment, were explained as due to creation and removal of lattice defects, predominantly those acting as acceptors.

This interpretation seems to be confirmed by the discovery of the effects of irradiation with high energy particles (α-particles, deuterons, electrons and fast neutrons).[7] Essentially identical effects have been observed for the conductivity changes of germanium after irradiation with any of these particles. At the beginning of the bombardment of *n*-type material the number of conduction electrons decreases linearly with increasing bombardment time. The resistivity passes through a maximum, corresponding roughly to the intrinsic resistivity of germanium, and thereafter decreases as the material becomes an increasingly good *p*-type conductor, as shown in Fig. 1. Bombardment of *p*-type germanium changes its conductivity monotonely toward a saturation value; the conductivity rises

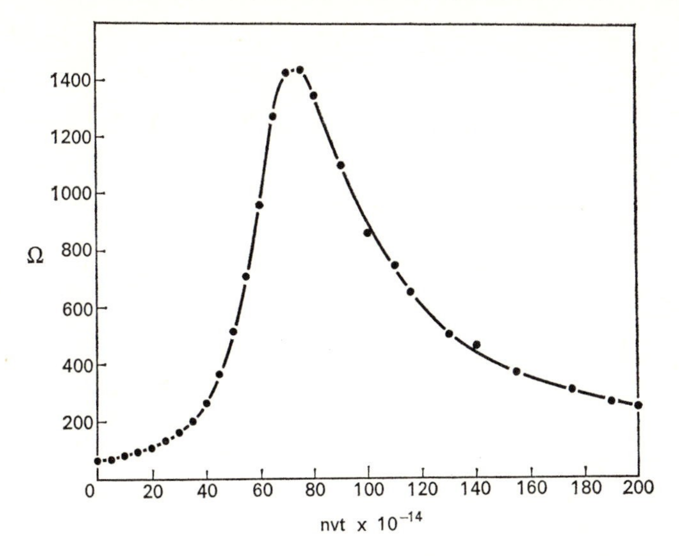

FIG. 1. Resistance of high resistivity n-type germanium bombarded by neutrons, as a function of integrated flux nvt of fast neutrons. Resistance measured at ambient pile temperature.

unless it is initially very high. (Compare Fig. 4, to be discussed later.)

A quantitative analysis[8] of the variation of conductivity of germanium with neutron bombardment (Fig. 2)[9] shows that the conductivity decreases at first at a rate corresponding to 3·2 carriers removed per incident neutron, approximately in agreement with calculations of G. E. Evans.[10] After the minimum conductivity is reached, Hall effect measurements indicate that the material is p-type, and the slope of the conductivity curve indicates the addition of about 0·7 holes per incident neutron at room temperature. The dashed curve of Fig. 2 is calculated assuming that the bombardment-introduced acceptors are equally effective as electron traps on the n- and p-type branch, that no annealing takes place, and that the mobility is unchanged by the imperfections introduced by bombardment.[11] The two curves are closely

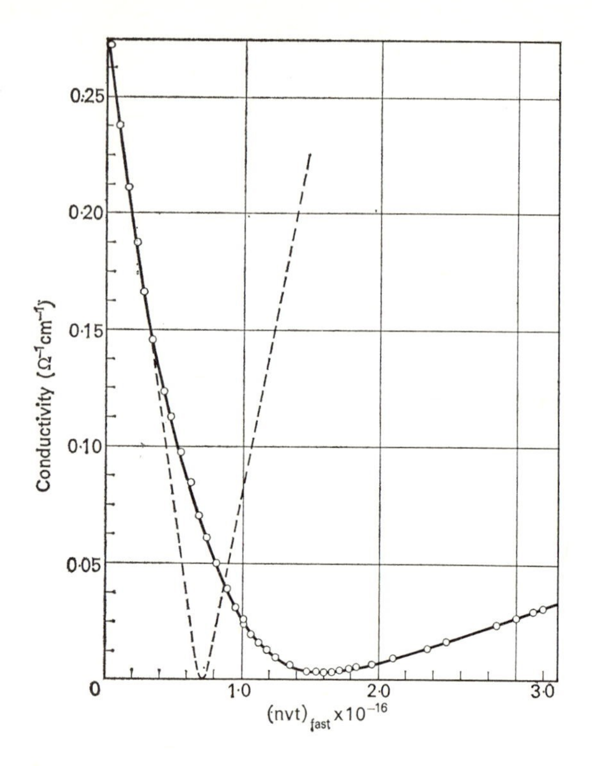

FIG. 2. Conductivity of *n*-type germanium bombarded by neutrons, as a function of integrated flux of fast neutrons. Exposure and measurement at ambient pile temperature.

analogous to conductiometric titration curves obtained on addition to an acid of an alkaline buffer and of a strong base, respectively. The difference in the magnitude of the slopes of the two branches of the observed curve indicates that the imperfections produced by bombardment are less effective as acceptors of electrons from the full band than as acceptors of electrons from the conduction band, and this implies that the activation energy of these acceptors is greater than that of chemical impurities, which are almost completely ionized at these temperatures. This is also confirmed by Hall effect and resistivity measurements of heat treated and quenched,[12] and bombarded samples.[5]

Recently Brattain and Pearson have measured and analyzed the effects of α-bombardment.[13] In the initial stages of the bombardment of *n*-type germanium, Brattain and Pearson observe the disappearance of 78 electrons per incident α-particle. Comparing this with the calculations of Seitz,[14] which indicate that there will be roughly 59 displaced Ge atoms per incident α-particle, they suggest that one acceptor or electron trap appears for each displaced atom, and that about one quarter of these acceptors contribute to the conductivity of the strongly bombarded *p*-type material.

We do not believe that so simple a picture of the process can be maintained. Bombardment produces not only vacancies in the lattice, which may act as acceptors, but also interstitial atoms, which can and must act as donators. If one pictures an interstitial Ge atom as a Ge^+ ion plus an electron embedded in a medium with the dielectric constant of germanium (16), one would estimate the binding energy for this electron as some $13\cdot54/16^2$ eV $= 0\cdot05$ eV. (More careful considerations, to be indicated later, indicate that the actual value may be somewhat different, but not greatly so.) Such an interstitial Ge atom will evidently lose an electron to any acceptor or low-lying electron trap in the forbidden band, and will thus tend to annul its effect in increasing the number of holes or decreasing the number of conduction electrons. One has then the question: how can bombardment-produced vacancies in the lattice serve as acceptors to produce hole conduction, when interstitial Ge atoms are simultaneously acting as donators?

In the case of silicon the behaviour on bombardment is rather different: (7a, c, e) the conductivity of bombarded silicon decreases as bombardment proceeds, whether it was originally of *n*- or *p*-type. One can describe this behavior as due to the simultaneous production of electron traps and hole traps, but a more detailed consideration of the situation is obviously demanded. In addition, one has a new question: why do these very similar materials, germanium and silicon, behave so very differently in this respect? This paper has the following purposes:

1. to show that vacancies in germanium and silicon may act as acceptors, and interstitial atoms may act as donators, not just for one electron, but for several;
2. to show that in these materials it is the second ionization energy of the acceptors and donators, rather than the first, that is critical for the behavior of the bombarded crystal;
3. to suggest that in germanium at least part of the interstitial atoms are doubly ionized, and part of the vacancies hold two electrons;
4. to indicate an interpretation of the different behaviors of germanium and silicon.

II. Substitutional Impurities and Interstitial Atoms as Donators

To begin, it may be desirable to give a precise definition of acceptors and donators. A donator is an imperfection which, in an electrically neutral and otherwise perfect crystal, and at 0°K, will cause an electron to appear either in the conduction band, or, more commonly, in a localized state in the forbidden band next below it. An acceptor is an imperfection which, in an electrically neutral and otherwise perfect crystal, and at 0°K, produces a hole in the highest "full" band, or, more commonly, an empty localized state in the forbidden band just above it.

In a crystal containing many imperfections, and at a temperature above 0°K, donators may increase electron conduction, or they may decrease hole conduction by donating electrons to fill holes in the "full" band (that is, by serving as "hole traps"). Similarly, acceptors may contribute to hole conduction by accepting an electron from the full band, or they may act as electron traps by accepting electrons from the conduction band.

It is well known that a substitutional impurity with one extra nuclear charge will serve as a donator. The attraction of the extra nuclear charge for electrons will perturb the band structure of the crystal downward in its vicinity. The corresponding extra electron, if placed in the conduction band, will be attracted to this region, and, at 0°K, will be trapped in a localized state with energy somewhat below the edge of the conduction band.[15] It is customary to compute this binding energy as that of an electron in the coulomb field of the perturbing nuclear charge, the interaction of these charges with the crystal being taken into account by introduction of

(a) the dielectric constant of the crystal, ϵ, and
(b) the effective mass m^* of the electron moving in the conduction band of the crystal. One usually takes m^* as the ordinary electronic mass.

Solution of the corresponding wave equation for the motion of the electron yields a 1s orbital which is not the complete wave function for the trapped electron, but is essentially the appropriate modulating factor for a periodic function characteristic of the crystal. The energy thus computed is a good approximation to the correct binding energy, provided only that the computed orbital is large compared to the diameter of a crystal cell. If this condition is not met, the concept of effective mass, which is strictly applicable only to an unperturbed crystal, ceases to be relevant; more important, the polarization of the crystal matrix becomes less effective in reducing the interaction of the nuclear charge and the electron, and the binding energy becomes correspondingly greater. In crystals like germanium and silicon the dielectric constant is very high, and the radius of the orbital of the trapped electron, which is $m/m^*\epsilon a_H$, is so large that this method of calculation should be reasonably accurate.

If the substitutional impurity has two extra nuclear charges, an electron in the conduction band will be more strongly trapped by it. Treating the two extra nuclear charges and the electron as a hydrogenic system, as just described, one will expect the energy of binding of this electron to be increased by a factor $Z^2_{\text{eff}} = 4$, and the radius of its orbital to be decreased by a factor Z_{eff} to $\frac{m}{2m^*}\epsilon a_H$.

But even in germanium, with $\epsilon = 16$, this radius is becoming comparable to the distance between Ge atoms, 4·6 a_H, and one must expect the binding energy of the electron to be considerably greater than is indicated by the hydrogenic model, and the radius of the actual orbital to be correspondingly less. For still larger excess nuclear charges the trapped electron would probably move in an orbital not greatly modified by the presence of the crystal matrix.

When the substitutional impurity with two extra nuclear charges has trapped one electron, the resulting ion can still attract and trap a second electron in a relatively large hydrogenic orbital corresponding to $Z_{eff} \simeq 1$. The binding energy for this second electron would be comparable to that for the first electron trapped by an impurity with one extra nuclear charge.[16]

To fully represent a substitutional impurity with two extra nuclear charges, one needs to introduce into the band picture two energy levels, lying below the edge of the conduction band by the two successive ionization energies discussed above. The lower of these levels corresponds to the trapping of an electron by an impurity with double excess charge to form an ion with single excess charge, the higher level to the trapping of an electron by this ion to form a lattice ion with normal charge. It is true that the second of these levels can never be occupied unless the first is occupied, in contrast to the usual situation in the band picture, where one treats each level as subject to occupation independently of all others. However, if these levels are far apart as compared to kT one can proceed exactly as if these ionization energies were associated with different impurities, since there will be no position of the ζ-level for which the Fermi distribution will indicate appreciable probability for occupancy of the upper level without also indicating very high probability for occupation of the lower level.

One can expect to apply a similar argument to the action of an interstitial atom as a donator. Consider a neutral interstitial Ge atom. An electron removed from this will be attracted by the resultant Ge^+ ion, and trapped in a localized state. If one can treat the Ge^+ ion as a point charge, one can estimate the trapping energy as 0·05 eV, as already indicated. The binding energy may be somewhat greater because of the size of the Ge^+ ion and the decreased effective dielectric constant near the interstitial nucleus, but it will at least be of this order. Next let us consider the Ge^+ ion by itself. An electron removed from this will be attracted by

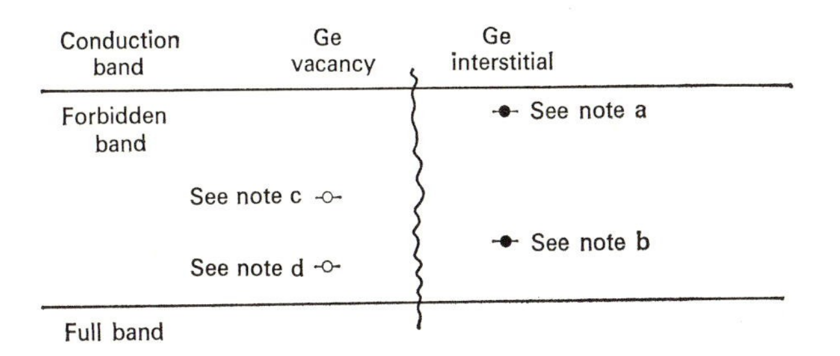

FIG. 3. Level diagram for vacancies and interstitials in germanium.
(a) If occupied, the interstitial atom is neutral.
(b) If not occupied, interstitial atom is doubly charged.
(c) If occupied, net charge of two electrons near vacancy.
(d) If not occupied, zero net charge near vacancy.

the resultant Ge^{++}, and will be trapped, with ionization energy certainly as large as $4 \times 0{\cdot}05$ eV, and probably considerably higher. Reasons will be given later for believing that this second ionization energy is of the order of the width of the forbidden band in germanium, 0·75 eV. The third and fourth ionization potentials of Ge will be very high, and the corresponding levels will lie among the full bands of the crystal, and will certainly be filled under equilibrium conditions. To represent an interstitial Ge atom in the band structure we will thus need to consider only two levels, lying below the conduction band by the first and second ionization energies of the Ge interstitial, and both filled at 0°K, as shown in Fig. 3.

III. Vacancies in Polar Crystals as Acceptors and Donators

By a simple inversion of the preceding argument it is easy to see that a substitutional impurity with nuclear charge decreased by 1 will tend to serve as an acceptor, trapping a hole in a state somewhat above the bottom of the full band, and accepting an electron from the full band when sufficient energy is supplied. Similarly, an impurity with nuclear charge decreased by 2 will serve as a double acceptor.

One might attempt to investigate the effect of a vacancy in a crystal by picturing the step-by-step reduction of a nuclear charge to zero, with removal of a corresponding number of electrons. The large number of steps required and the magnitude of the resulting perturbation make it difficult to draw secure conclusions from such considerations, and other methods must be employed. One of these will be illustrated for the case of polar crystals, in which negative ion vacancies can serve as donators, and positive ion vacancies as acceptors.

In the usual way, let us picture the formation of a KCl crystal by gradual reduction of the ionic separations in a regular array of K^+ and Cl^- ions. The full shells of the ions give rise to full bands in the crystal, which is an insulator because of the great width of the forbidden band just above these full bands.

Now if the single Cl^- ion is omitted from this array, the full shells of electrons again go over into full bands. There will, of course, be two fewer states in each band arising from the Cl^- orbitals, but, since the corresponding electrons are absent with the missing Cl^- ion, the band structure of the resulting system will consist entirely of bands that are completely filled and bands that are completely empty. The regularity of the crystal is disturbed by the missing negative charge, and energy levels corresponding to localized states may be split out, downward, from the full and from the empty bands. The states split out from the full bands will, however, continue to be filled, and those split out from the empty bands will continue to be empty, provided, of course, that they are not depressed into the full bands. This is not to be expected, in view of the great width of the intervening forbidden band.

Now let an electron be returned to the crystal with Cl^- vacancy, in order to restore its electrical neutrality. It is well known that the excess positive charge around the vacancy will trap this electron, to form an *F*-center. The *F*-center will give up the electron to the conduction band if sufficient energy is supplied, thus acting as a donator with rather high activation energy. We see, then, that a Cl vacancy in KCl will act as a donator.

In the same way, one can picture the formation of a KCl crystal from an array of K^+ and Cl^- ions from which one K^+ ion is missing. Again the closed shells of these ions go over into completely filled bands in the resulting system. Absence of the positive ion may cause energy levels, corresponding to states localized about this vacancy, to split out, upward, from the full and from the empty bands. In particular, one will expect such a level to appear in the forbidden band above the top full band; like the band from which it is derived, it will be filled at 0°K. To restore electrical neutrality of the system, a hole can now be introduced into the full band. This will be attracted to the region of excess negative charge and will be trapped there, as the electron is at an *F*-center; in other terms, the hole will be filled by the electron from the localized state in the forbidden band, which will thus become empty. This K^+ vacancy with localized empty state, or trapped hole, is a *V*-center.[17] It will accept an electron when sufficient energy is supplied (absorption of light in the ultraviolet), and will thus act as an acceptor with very high activation energy.

In a crystal containing both *F*-centers and *V*-centers the energy levels of the *F*-centers lie lower and continue to be filled while the *V*-center states remain empty.

IV. Vacancies in Germanium as Acceptors

The discussion of polar crystals in the preceding section is simplified by the fact that the perturbation acting between the ions in the lattice is relatively small, and does not cause crossing or overlapping of bands as the ions are brought together. In germanium and silicon the situation is known to be very different;[18] it will here be described very briefly.

Let a germanium crystal be pictured as formed by gradually bringing together a regular array of Ge atoms. The closed inner shells of the atoms give rise to full bands in the crystal. For large atomic separations the $4s$, $4p$, $4d$ orbitals of the atoms go over into corresponding bands. As the atomic separations are reduced there is an overlapping and crossing of these bands, until, at the physically existing nuclear separation, the primary factor in determining the energy of the bands is not the nature of the atomic orbitals involved, but the symmetry of the wave functions with respect to the cell boundaries. The highest full bands in Ge are bands for which the band-edge functions are symmetrical about the cell boundaries; above the forbidden band lie empty bands for which the band-edge wave functions are antisymmetric about the cell boundaries.

Now let a Ge^{++++} ion be removed from this array. As concerns the lowest-lying full bands of the crystal, which arise from the closed shells of the Ge atoms, it is evident that the number of states in these bands will be reduced by a number equal to that of the electrons removed with the ion; for instance, the K-bands of the perturbed crystal will contain two fewer states than before. These lowest bands need not be considered further in counting the number of states available and the number of electrons to fill them. So far as the remaining bands of the crystals are concerned, removal of the ion can be considered merely as removal of four units of positive charge from the site of the ion. The band structure of these valence-electron bands will be perturbed upward at the vacancy, and localized states will be split off, upward, from the full valence-electron bands. If the removal of the ionic charge is thought of as adiabatic, the states thus split off from the filled bands will continue to be filled; the later redistribution of electrons among the states will be considered separately.

The removal of four units of positive charge is so large a perturbation, and the band structure of germanium is so complex, that one can not forsee with certainty how many localized states will split off upward from the full bands into the forbidden band, and beyond. We shall therefore need to consider several possibilities.

First, let us suppose that four localized states are thus split off. So long as these states are filled, the density of electrons in the vacancy will be as great as it was before the Ge^{++++} ion was removed. If now four holes are added to the full band, to restore neutrality of the crystal, they will be attracted to this region of excess negative charge, and will there be filled by electrons falling into them from the four localized states, which will thus become empty. The vacancy will then lack not only a Ge^{++++} ion, but four electronic charges. An electron can be forced into this vacancy, a hole removed from it, by supplying a sufficient activation energy. To bring a second electron into this vacancy, remove a second hole, will require considerably greater energy, because of the strong coulomb repulsion of the electrons in the vacancy. To bring a third and fourth electron into this vacancy would require many electron volts of energy, because of this repulsion. Considering this process from the hole point of view, one may speak of the first, second, and higher ionization potentials for the removal of holes from the vacancy. Concerning those ionization potentials, this picture suggests only that the third and fourth ionization potentials will be very high, and of no interest in the present connection. The vacancy can then be represented in the band picture by energy levels lying above the top of the full band by the first and second hole ionization energies. Occupation of only the lower state will correspond to the presence of one electron in the vacancy, occupation of both to the presence of two electrons. Both states will be empty at 0°K, and will act as acceptors, according to our definition, if they lie in the forbidden band.

It is not clear, however, that removal of a Ge^{++++} ion will split off from the full bands as many as four filled and strongly localized states. Suppose that only three such states are split off. When four holes are added to the full band, to restore electrical neutrality, three of these will be filled by electrons from the three localized states, which thus become empty. There will continue to be an excess of one electronic charge in the neighbourhood of the vacancy, to which the one remaining hole will be attracted. One can expect to treat this system by the hydrogenic model, as a fixed negative charge plus a hole of effective mass m^*_h. At 0°K one will then expect the hole to be trapped by the vacancy-plus-electron, in an orbital with radius about $\frac{m}{m^*_h}\epsilon a_H$, with a dissociation energy of the order of 0·05 eV. (This system would have the character of an exciton, with its electron trapped at the vacancy.) A second electron could be introduced into the vacancy by supplying sufficient energy, corresponding roughly to the distance of the first localized state above the full band; the amount of energy required is not indicated by this argument. A third and fourth electron could also be introduced, but the energy required would be so great, because of the coulomb repulsion of the electrons, that these states need not be considered here. Again one can represent the vacancy in the band picture by two levels, one lying above the edge of the full band by the trapping energy of the hole, some 0·05 eV, the second by the energy required to introduce an additional electron into the vacancy. At 0°K both states will be empty, as illustrated in Fig. 3. It is to be emphasized that in this figure the vacancy and interstitial atom are not supposed to be present simultaneously.

The difference between the situations pictured in the two preceding paragraphs is the following. In the first, it is assumed that the crystalline field with a Ge atom missing is such that the electronic density in the vacancy will be very low at 0°K, and the electronic density in the nearby parts of the crystal will be little changed by the vacancy. Excitation energies for the acceptor states are not computable by the hydrogenic model, being determined by the details of the crystalline field, and they may be high. In the second picture we have an electron held near the vacancy by the crystalline field and a hole attracted to the electron; at 0°K there will be an excess of negative charge near the vacancy, and a deficit of negative charge in a surrounding region of radius about $\frac{m}{m^*}\epsilon a_H$, due to the trapped hole. The first hole ionization energy can here be computed by the hydrogenic model, since it is determined by a coulomb interaction, but the second hole ionization energy is determined by details of the crystalline field, and can not be treated in the same way.

Extending the second picture, one may suppose that only two localized states are split off from the full band when the Ge^{++++} is removed. Then at 0°K two electrons will be held near the vacancy by the crystalline field, and two holes will be trapped in the neighbourhood by the coulomb attraction of the electrons. The first two hole ionization energies would then be computable by hydrogen- or helium-like models, perhaps with Z_{eff} less than 1 or 2, since the central negative charge is not localized at a point. Indeed, one can expect these models to be more satisfactory than in the case of interstitial Ge atoms, since the central attracting charge, being less concentrated, is less likely to lead to strongly bound orbitals of small radius, for which one can not treat the crystal matrix simply as a dielectric medium. As before, the third and fourth hole ionization energies would certainly lie so high as to be of no interest.

The choice to be made between these alternative pictures is in part suggested by the previously given description of the band picture of germanium, and in part by experimental results. The symmetry of the crystal orbitals, which is so important in germanium and silicon, can be maintained only if they extend through the vacancy—that is, only if the normal quota of electrons remains in the vacancy. This is certainly not the case, because of the large coulomb repulsion of the electrons, but the tendency in this direction may very easily result in the trapping of one electron near this vacancy, and possibly that of a second. It therefore appears that the first hole ionization energy should be computable by the hydrogenic model, and possibly even the second; to decide the latter point one can at present only refer to experiment. It is quite conceivable that in this respect there should be a marked difference between germanium and silicon, and it will be suggested later that this is indeed the case.

V. Effect of Bombardment of Germanium

We have now to discuss the effect of nuclear bombardment of germanium, considered as the production of equal numbers of isolated lattice vacancies and isolated interstitial atoms. It must, however, be pointed out that any tendency of vacancies or interstitials to form clusters would affect the quantitative aspects of the argument.[19]

Figure 4 illustrates[20] what happens to the ζ-level in a crystal when bombardment adds donators and acceptors in equal numbers. It is assumed that all donator states have the same energy, and all acceptor states the same energy. Different concentrations N_{c+} of chemical donators or concentrations N_{c-} of chemical acceptors are assumed to be present in the material before bombardment. Whatever the character and number of the impurities originally present, the donators and acceptors added by bombardment will eventually become dominant, and the ζ-level will approach a value determined by these imperfections only. This limit will actually lie midway between the donator and acceptor levels, this corresponding to the fact that, in the limit, for each electron absent from an acceptor level there must be one present in a donator level. It will be noted that this figure corresponds qualitatively to the observed behavior of Ge under bombardment; n-type material is converted into p-type, and the conductivity of p-type material is increased (if it is not very high in the beginning) as the ζ-level draws nearer to the edge of the filled band.

We have already seen that in bombarded germanium the situation is somewhat more complicated than this. Let us suppose that the interstitials and vacancies of Fig. 3 are present in equal numbers. For each vacancy-interstitial pair there are four available states, and two electrons to go into them. As the number of such pairs increases the ζ-level will approach a limit midway between the second and third of these levels, provided only that kT is

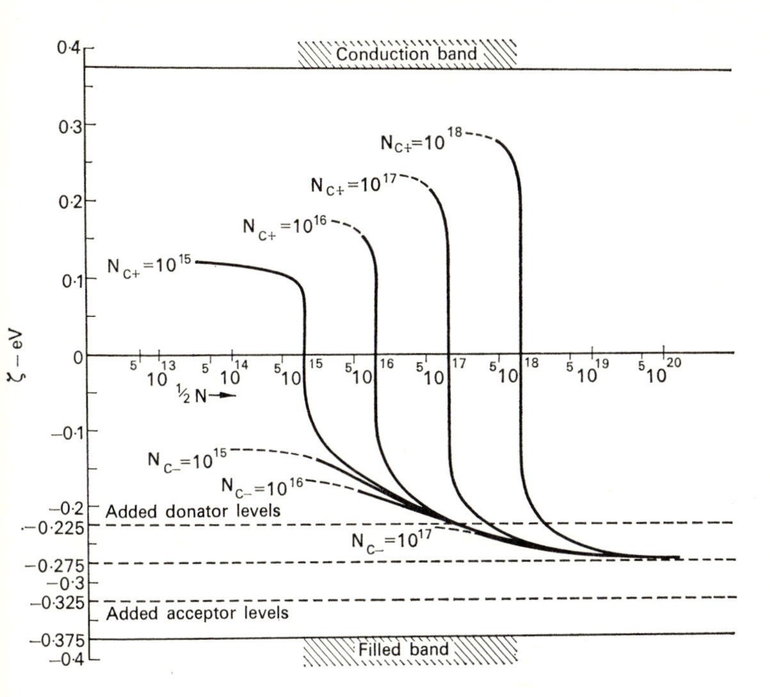

FIG. 4. Change in ζ-level against number $N/2$ of added donator or acceptor levels, or various initial concentrations N_{c+} of chemical donators or N_{c-} of chemical acceptors.

small compared to the energy separations of the two highest and of the two lowest levels. (In this case the second level must be occupied in as many cases as the third level is empty.)

In the case of bombarded Ge, conductivity observations indicate that the ζ-level approaches a limit near the bottom of the forbidden band. It follows that the three lowest levels in Fig. 3 all lie fairly close to or below the edge of the forbidden band. This is consistent with a second ionization energy for the interstitial atoms that is near the width of the forbidden band, 0·75 eV, or greater than this, but not much smaller. It indicates also that the second hole ionization energy for the vacancy is not large, probably less than

0·20 eV. This suggests that the second hole ionization energy should be computable using a hydrogenic model, with Z_{eff} perhaps less than 2.

Hall effect measurements[5] of bombarded samples at low T show a curvature in the plot of Hall coefficient against $1/T$, which indicates that there are several energy levels near the edge of the forbidden band. If these levels correspond to successive hole ionization energies of a vacancy, they can not be treated as subject to independent occupation, in the way previously assumed. It seems more likely, however, that the levels in question are the first ionization level of the vacancies and the second ionization level of the interstitials, with the second ionization level for the vacancies lying somewhat higher. It thus appears that the second ionization energy of an interstitial Ge atom is a little less than 0·75 eV.

The above considerations make it clear that in strongly bombarded germanium all interstitial atoms are at least singly ionized. They suggest also that the position of the ζ-level is determined by an equilibrium between single and doubly-ionized interstitials, on the one hand, and singly and doubly-ionized vacancies, on the other. Details of the arrangement of the levels, and deduction of the relative number of singly and doubly ionized imperfections, can be settled only by a detailed analysis of the variation of Hall constant and conductivity with bombardment.

Confirmation of this picture is provided by closer inspection of Fig. 2. According to this model, each vacancy-interstitial pair provides two vacant levels that can trap two electrons from the conduction band. The lower of these levels will lie far below the conduction band, and will almost certainly be filled, but the higher level, corresponding to the first ionization energy of a Ge interstitial, will be filled with a probability that depends on its position, on the temperature, and on the initial position of the ζ-level. When high conductivity germanium is bombarded there should be about two conduction electrons trapped per displaced Ge atom, or vacancy-interstitial pair: for material with lower conductivity the value should be somewhat less. This conclusion seems to agree with the observations and calculations mentioned in section I as well as does the hypothesis of Brattain and Pearson, and as well as can be expected from the accuracy of Seitz's calculations. But beyond this, it should be noted that the resistivity of the material will not become really high until the conduction electrons are all trapped in the lower-lying levels, and it is not necessary for the highest, easily ionized, interstitial level to act as a trapping level. This will happen only after bombardment has displaced twice as many atoms as would be required if all vacant levels were fully effective as traps; that is, it will occur only after twice as long a bombardment as one would expect from linear extrapolation of the initial part of the curve of conductivity against bombardment, if the initial conductivity is high. It is evident from Fig. 2 that the minimum conductivity is, indeed, reached shortly after this stage of the bombardment. "Healing" of the bombardment-produced imperfections will also tend to delay passage through minimum conductivity, but it does not appear to be the primary factor here.

VI. Effects of Bombardment of Silicon

The effects of nucleon bombardment of silicon will be indicated briefly, since the observations now available are too incomplete, and in part too tentative, to justify any attempt at complete analysis.

One must expect the lower dielectric constant of silicon, 12, to lead to ionization energies of interstitial atoms which are higher than the corresponding ones in germanium. The fractional increase will certainly be greater for the first ionization energy than for the others, which correspond to the ionization of more compact positive ions. One can thus expect the second ionization energy of silicon interstitials to be somewhat greater than that in germanium $\sim 0 \cdot 75$ eV; at any rate, the corresponding level will lie well below the middle of the forbidden band, despite its greater width, $1 \cdot 2$ eV.

Neutron bombardment of samples of silicon of good but not great purity (resistivity ~ 1 ohm-cm) rapidly raises the resistivity of both *n*- and *p*-type materials; it is evident that the limiting position of the ζ-level must be far from the edge of the full band. It follows that the second hole ionization energy of silicon vacancies must be much greater than that in germanium. Now, bombarded *p*-type silicon develops a conductivity with activation energy roughly 0·65 eV.[(6b)] Since the bombarded sample appears to have remained *p*-type, one must attribute this conductivity to the second hole ionization of the vacancies; that is, the second-lowest vacancy level in silicon must be roughly 0·65 eV above the edge of the full band, and well above the second highest interstitial level. It follows that in bombarded silicon containing large and equal numbers of vacancies and interstitials these imperfections will almost all have single negative and positive charges, respectively.

Bombarded silicon, whether of *n*- or *p*-type, develops an infrared absorption band at a wavelength corresponding to an excitation energy of 0·68 eV.[21] This absorption band becomes increasingly sharp at lower temperatures,[22] and no photoconductivity has been found to be associated with it. It, therefore, appears to be due to a transition between discrete levels, probably to excitation rather than ionization of the interstitial Si^+ ions. If the second ionization energy of the Si interstitials is less than 1.2 eV (as is indicated by the rising resistivity of bombarded *p*-type silicon), ejection of electrons from Si^+ ions into the conduction band will give rise to an infrared absorption somewhat to the long-wave side of the absorption edge in the pure material; the question of how far the shift in the apparent band edge, as observed by M. Becker,[23] can be attributed to this absorption will have to be studied in detail. It thus appears that the second ionization energy of silicon interstitials is around 1 eV.

The most striking difference between the energy levels in bombarded germanium and silicon, as indicated by this analysis, is the much greater second hole ionization energy of silicon vacancies. Such a difference must be attributed to differences in the crystalline fields of these materials, as discussed in section IV, and is not to be explained by hydrogenic models.

NOTES

[1] Work supported in part by AEC.

[2] J. FRENKEL, *Z. Phys.* **35,** 652 (1926). C. WAGNER and W. SCHOTTKY, *Z. phys. Chem.* (B) **11,** 163 (1930); W. JOST, *J. Chem. Phys.* **1,** 466 (1933); W. SCHOTTKY, *Z. Elektrochem. angew. phy. Chem.* **45,** 33–72 (1939), particularly on the statistics and thermodynamics of disordering in crystals. See also N. F. MOTT and R. W. GURNEY, *Electronic Processes in Ionic Crystals*, pp. 26–63 (1948).

[3] K. LARK-HOROVITZ, Final Report NDRC 14–585, 1942–45 for a summary of the early work at Purdue University, and W. SHOCKLEY, *Electrons and Holes in Semiconductors*, D. Van Nostrand Co., New York, 1950, for a summary of work at Bell Telephone Laboratories.

[4] G. L. PEARSON, J. D. STRUTHERS, H. C. THEUERER, *Phys. Rev.* **75,** 344 (1949).

[5] See K. LARK-HOROVITZ, *Electr. Engng.* **68,** 1047–1056 (1949) and J. CLELAND, K. LARK-HOROVITZ, J. C. PIGG, *Phys. Rev.* **78,** 814 (1950).

[6] NDRC 14–585, p. 24. For more recent investigations, see W. E. TAYLOR, Ph.D. thesis, Purdue University, 1950 and *J. Met.* (to be published soon). H. C. THEUERER and J. H. SCAFF, *J. Met.* **189,** 59 (1951).

[7] For a summary of this work see K. LARK-HOROVITZ, "Nucleon-irradiated semiconductors", Conference on Properties of Semiconducting Materials, Reading, England, July, 1950, and in detail: (a) K. LARK-HOROVITZ, E. BLEULER, R. DAVIS, D. TENDAM, *Phys. Rev.* **73,** 1256 (1948). (b) R. E. DAVIS, K. LARK-HOROVITZ, Signal Corps Progress Report, Aug.–Oct. 1948. (c) R. E. DAVIS, W. E. JOHNSON, K. LARK-HOROVITZ, S. SIEGEL, *Phys. Rev.* **74,** 1255 (1948). (d) R. E. DAVIS, W. E. JOHNSON, K. LARK-HOROVITZ, S. SIEGEL, AECD Report 2054, June 1948. (e) W. E. JOHNSON, K. LARK-HOROVITZ, *Phys. Rev.* **76,** 442 (1949). (f) E. KLONTZ, K. LARK-HOROVITZ, Signal Corps Progress Report, Aug.–Oct. 1948 and December 1949 and following. (g) J. H. CRAWFORD and K. LARK-HOROVITZ, *Phys. Rev.* **78,** 815 (1950). (h) J. H. CRAWFORD and K. LARK-HOROVITZ, *Phys. Rev.* **79,** 889 (1950). (i) E. KLONTZ and K. LARK-HOROVITZ, *Bull Amer. Phys. Soc.* **26,** 7 (Mar. 1951). (j) V. A. JOHNSON and K. LARK-HOROVITZ, *Bull. Amer. Phys. Soc.* **26,** 7 (Mar. 1951). (k) J. W. CLELAND, J. H. CRAWFORD, K. LARK-HOROVITZ, J. C. PIGG, *Bull. Amer. Phys. Soc.* **26,** 7 (1951).

[8] J. H. CRAWFORD, JR. and K. LARK-HOROVITZ, *Phys. Rev.* **78,** 815 (1950).

[9] From a manuscript of J. W. CLELAND, J. H. CRAWFORD, JR., K. LARK-HOROVITZ, J. C. PIGG, F. W. YOUNG, *Phys. Rev.* **83,** 312 (1951).

[10] G. E. EVANS, *Phys. Rev.* (to be published).

[11] Experiments at the Oak Ridge National Laboratory indicate that acceptors and donators introduced by nuclear transmutations are equally effective, but this is not necessarily true of lattice imperfections.

[12] A. E. MIDDLETON, Ph.D. thesis, Purdue University, 1945.

[13] W. H. BRATTAIN and G. L. PEARSON, *Phys. Rev.* **80,** 846 (1950).

[14] F. SEITZ, *Discuss. Faraday Soc.* **5,** 217 (1949).

[15] See, for instance, H. M. JAMES, *Phys. Rev.* **76,** 1611 (1949).

[16] It might appear more appropriate to consider the two trapped electrons plus the added nuclear charge as forming a He-like structure embedded in a dielectric. This would lead one to a somewhat higher estimate of the first ionization energy for the system. It appears to us, however, that the considerations employed here are more appropriate. In treating the atom He one gets a much more accurate wave function by considering the electrons as moving in non-equivalent orbitals than by treating the orbitals as equivalent. (See, for instance, *Introduction to Quantum Mechanics*, by Pauling and Wilson, Table 29—1, p. 224, McGraw-Hill Book Co., New York, 1935.) This non-equivalence would be greatly exaggerated in the present case due to the rapid decrease of dielectric shielding at small *r*. It thus appears reasonable, as a first approximation, to neglect the effect of the second trapped electron on the orbital describing the motion of the first one.

[17] F. SEITZ, *Rev. Mod. Physics* **18,** 384 (1946). But see also F. SEITZ, *Phys. Rev.* **79,** 529 (1950).

[18] J. F. MULLANEY, *Phys. Rev.* **66,** 326 (1944).

[19] F. SEITZ, *Phys. Rev.* **79,** 890 (1950). K. LARK-HOROVITZ, Reading Conference Report, Note 7.

[20] K. LARK-HOROVITZ, Reading Conference on Properties of Semiconducting Materials, Appendix II, by G. W. LEHMAN and H. M. JAMES.

[21] K. LARK-HOROVITZ, M. BECKER, R. E. DAVIS and H. Y. FAN, *Phys. Rev.* **78,** 334 (1950).

[22] M. BECKER, Ph.D. thesis, Purdue University, 1951.

[23] M. BECKER, Signal Corps Progress Report, Dec. 1950–Feb. 28, 1951.

17

Fast Neutron Bombardment of Germanium†

The Effect of Fast Neutron Bombardment on the Electrical Properties of Germanium

ABSTRACT

Lattice disorder in bulk Ge produced by collisions of fast neutrons with lattice atoms introduces a net excess of electron traps or acceptor states, which appear to have an energy distribution in the forbidden band. Bombardment of *N*-type Ge causes the conductivity to decrease, initially at a uniform rate, to a minimum value and then to increase. Examination of the data shows that initially about 3·2 conduction electrons are removed per incident fast neutron. The minimum value of the conductivity is higher than the value calculated assuming complete homogeneity and thermal equilibrium. Hall effect measurements prove that after the minimum is passed the material has been converted to *P*-type by fast neutron bombardment. The conductivity of *P*-type Ge increases with bombardment. The rate of increase decreases monotonically indicating an approach to saturation. The initial rate of carrier introduction in *P*-type Ge appears to be temperature dependent, being greater at higher temperatures (~ 0·8 carrier per incident neutron at 30°C), and smaller than the rate of carrier removal for *N*-type Ge which is apparently temperature independent in the temperature range investigated (− 79°C to 45°C). The effect of lattice disordering on the electrical properties of Ge may be removed by careful vacuum annealing at 450°C while a portion of this effect readily anneals at room temperature.

I. INTRODUCTION

The effect of high energy nucleon bombardment on the electrical properties of Ge and Si was first investigated by Lark-Horovitz and co-workers[1] using 10-MeV deuterons and 20-MeV alpha-particles from the Purdue cyclotron. They also investigated the

† *Phys. Rev.* **83,** 312 (1951) and *ibid.* **84,** 861 (1951), with J. W. Cleland, J. H. Crawford, Jr., J. C. Pigg, and F. W. Young, Jr.

effects of neutrons from the Be–D reaction and alpha-particles from a polonium source. The Purdue group in collaboration with Johnson and Siegel[2] investigated the effect of fast neutrons by irradiating bulk Ge and Ge-diodes in the Oak Ridge nuclear reactor. This work has been extended by the authors, and the present paper is intended to summarize the results of investigations to date on the effect of fast neutrons on the ohmic properties of bulk Ge. More recently Brattain and Pearson[3] have also studied the effect of alpha-particles from a polonium source.

When *P*-type Ge is bombarded with high energy nucleons the conductivity increases monotonically with bombardment. The conductivity of *N*-type Ge, on the other hand, first decreases, passes through a minimum and then increases with further bombardment. Hall coefficient measurements prove that the bombarded *N*-type material has been converted to *P*-type. The only difference between the effects of charged particles and neutrons is in the distribution of the effect in a massive target. The bombardment effects induced by charged particles are limited to their range in the material (about 0·2 mm for 10-MeV deuterons,[1] and only 0·019 mm for 5·3-MeV polonium alpha-particles).[3] In order to achieve complete penetration with charged particles, samples thinner than the effective range must be used, and even then, because of range-energy relationships, the effect is not uniform throughout the thickness of the target. On the other hand, because of their relatively large mean free path, fast neutrons produce effects which are uniformly distributed throughout a bulk target of relatively large dimensions. The advantages of the use of fast neutrons in a study of the effect of nucleon bombardment on the electrical properties of Ge are therefore obvious. It should be mentioned, however, that there are disadvantages associated with irradiations in a nuclear reactor. Here one is dealing with a wide neutron energy spectrum, i.e., from thermal energies up to the MeV range, and high intensity β- and γ-radiation as well. Thus the effect of nuclear reactions induced by thermal neutrons must be considered as must also the effect of β- and γ-radiations.

The bombardment produced change in the properties of Ge is much too large to be explained on the basis of impurities introduced by nuclear reactions. The number of impurity atoms produced by transmutations with charged particles is extremely small although this does play an important part in the case of pile bombardment.[4] Lattice disordering and lattice displacements are caused by elastic collisions of the bombarding particles with lattice atoms and by secondary collisions of struck atoms or "knock-ons" with those of the lattice. This disordering process produces a net excess concentration of electron traps or acceptors. Hole traps may also be produced but in the case of Ge these are not nearly so efficient as the electron traps, else either *P*- or *N*-type Ge would tend toward an intrinsic semiconductor with bombardment.[4a] These traps or acceptors neutralize donors in *N*-type Ge, thus decreasing the current carrier concentration, and in the case of *P*-type Ge, if they are sufficiently deep lying, they may augment the concentration of positive carriers in the filled band. All of the electrical effects of lattice disordering may be removed by annealing the crystals at about 450°C.

The behavior accompanying bombardment is analogous in many ways to that observed with appropriate heat treatment. If *N*-type Ge is quenched from temperatures near the melting point ($\sim$ 800°C), *P*-type Ge is produced.[5] This conversion of type is presumably due to "frozen-in" Frenkel type lattice defects which act as acceptors. By annealing at 450°C and cooling slowly the original *N*-type character is restored.

II. Theoretical Discussion

When high energy particles are stopped by crystalline solids the primary particle loses its energy by two types of interactions with the lattice: (1) elastic energy losses by direct momentum transfer to atomic cores, thereby producing displaced atoms, and (2) inelastic energy losses in which electrons are excited by charge interaction. The latter type, of course, applies only to the interaction of charged particles. The struck atom (secondary particle or "knock-on") in turn also loses its energy in a similar manner until all of the energy imparted to the lattice by the primary particle is either dissipated or stored. The inelastic energy losses appear exclusively as heat in conductors but in the case of insulators some of this energy may be stored by trapped electrons (F-centers, etc.). Most of the elastic energy losses, however, are stored by the lattice in the form of displaced atoms. Seitz[6] has treated this problem theoretically and has calculated the number of displaced atoms to be expected from the interaction of various particles of specified energy with a number of materials.

In order to understand the behavior of the conductivity of Ge during fast particle bombardment one must know the effect of these displaced atoms on both the concentration of current carriers and on their mobility. We shall assume that the lattice disorder may be resolved more or less distinctly into two classes: (1) lattice vacancies and interstitial atoms in regions of small amounts of disorder and (2) complexes or clusters of disorder of the type considered by Vand[7] in regions of concentrated damage. In view of the similarity in the behaviour of Ge and Si on quenching from high temperature[5] to that observed on particle bombardment, one would expect the first class of disorder to produce hole and electron traps or donors and acceptors. Such centers of disorder will have associated with them localized charges which will scatter carriers, thereby decreasing the mobility. The role of the second class is not so clear. These may also be able to produce carriers. Also, certainly, damage centers of this type will scatter conduction electrons and holes.

The introduction of vacant states[7a] deep in the forbidden band of an *N*-type semiconductor will cause the Fermi level ζ (electronic chemical potential) to be lowered toward the filled band. If these traps are distributed in energy, as experiment seems to indicate, initially they will all remove electrons from the conduction band and, of course, the original donor level. After nearly all of the electrons are removed, further introduction will cause a redistribution of the trapped electrons to vacant states of lower energy, until essentially all of the electrons occupy the lower-lying states. Only after this redistribution has taken place can the low-lying vacant states begin to act as acceptors, causing the material to become *P*-type. Consequently, a constant rate of introduction of a distribution of low-lying vacant states may be visualized as causing (1) a rapid lowering of ζ toward the center of the forbidden band corresponding to the initial removal of electrons from the conduction band, (2) a gradual lowering of ζ across the center of the forbidden band corresponding to the redistribution of electrons to traps of lower energy and transition to *P*-type, and (3) a rapid, though not so precipitous as in (1), depression of ζ toward some limiting position near the top of the filled band. Actual calculations of this type have been carried out for the case in which acceptors of a single energy are added.[8]

Addition of acceptors to *P*-type Ge has the same effect as phase (3) above in the case of *N*-type Ge. High resistivity *P*-type Ge is readily affected, since in this situation ζ is well above the top of the filled band. In low resistivity *P*-type Ge, however, the Fermi level may already lie below the limiting position mentioned above. If such is the case, the only effect produced by bombardment on the conductivity will be through a decrease in the mobility. It should be emphasized that the increase in hole concentration in *P*-type Ge is less than the decrease in electron concentration for a given period of irradiation. This is expected since only that portion of acceptors which are thermally ionized are effective in increasing the hole concentration.

With regard to mobility any departure from the periodic potential of the lattice will tend to shorten the mean free path of a carrier because of additional scattering. The effect of localized charge in the lattice has been extensively investigated both theoretically[9] and experimentally.[10] Also a treatment of neutral scattering has been reported[11] more recently. Except for high concentrations of scattering centers, impurity and defect scattering is usually subordinate to scattering from lattice vibrations at moderate temperatures, the former being more important at low temperatures. Consequently, no effect on mobility is expected until the concentration of bombardment introduced defects becomes comparable to or greater than the initial impurity concentration at the exposure temperatures used in these experiments.

The minimum conductivity observed during conversion of N-type Ge to P-type by fast particle bombardment may be examined from the simple model of semiconductors. The general equation for the conductivity of a semiconductor capable of exhibiting intrinsic behavior is

$$\sigma = e\mu_e n_e + e\mu_h n_h, \tag{1}$$

where e is the electronic charge, μ is the mobility, n the carrier concentration and the subscript refers to the type of carrier. For carrier concentrations considered here classical statistics are valid. Thus for *thermal equilibrium* the carriers obey the relation

$$n_e n_h = \kappa(T) = 4h^{-6}(2\pi \bar{m}^* kT)^3 \exp(-\Delta\epsilon_g/kT), \tag{2}$$

where $\bar{m}^*$ is the geometric mean of the effective masses of the carriers and $\Delta\epsilon_g$ is the energy gap between the highest filled band and the conduction band. The best value of $\kappa(T)$ obtained from measured properties of Ge in the intrinsic range is given by[12]

$$\kappa(T) = 5{\cdot}15 \times 10^{31} T^3 e^{-8630/T}.$$

By using Eq. (2) to write Eq. (1) in terms of only one carrier concentration and minimizing the conductivity with respect to this carrier one finds that the carrier concentrations at the minimum are

$$n_e = (\kappa/c)^{\frac{1}{2}},$$
$$n_h = (\kappa c)^{\frac{1}{2}}, \tag{3}$$

where c is the ratio of electron mobility to hole mobility. Substituting these values into Eq. (1) gives for the minimum conductivity

$$\sigma_{\min} = 2e\mu_e(\kappa/c)^{\frac{1}{2}}. \tag{4}$$

It is evident from Eq. (3) that the conductivity minimum lies at the intrinsic concentration ($n_e = n_h$) only if $c = 1$. For Ge, $c \sim 1{\cdot}5$. Therefore, the material has already become P-type before the minimum is reached. The general expression for the Hall coefficient is given by

$$R = (3\pi/8e)(n_h\mu_h^2 - n_e\mu_e^2)/(n_h\mu_h + n_e\mu_e)^2. \tag{5}$$

Substitution in Eq. (5) of the carrier concentrations from Eq. (3) gives

$$R_{\min} = (3\pi/32e)(1 - c)/(\kappa c)^{\frac{1}{2}}. \tag{6}$$

Hence it is evident that the Hall coefficient is negative even though at the conductivity minimum the material is P-type. Similar examination of the thermoelectric power[13] shows that the Seebeck coefficient is given by

$$\theta_{\min} = -(k/5e)\ln c \tag{7}$$

which is also negative even though the material is P-type.

III. The Effect of Fast Neutrons on N-type Ge

A typical room temperature conductivity vs. bombardment curve for N-type Ge is shown in Fig. 1. The measurements were taken in-pile during bombardment. Initially the slope is quite linear, the conductivity decreasing at a constant rate with bombardment. As the minimum is approached the slope becomes less and less negative, producing a wide, flat minimum. After the minimum the conductivity increases much more slowly than the initial rate of decrease. By assuming that annealing of lattice disorder and changes in the electron mobility are negligible during the initial part of the bombardment, the average net number of electron traps produced per incident neutron K for the fast neutron spectrum which exists in the Oak Ridge reactor may be calculated from the initial linear portion of the curve. The initial mobility of the samples is known from Hall coefficient and resistivity measurements and K is readily calculated by means of the relation

$$K = dn_e/d(nvt)_{\text{fast}} = (1/e\mu_e)d\sigma/d(nvt)_{\text{fast}}. \tag{8}$$

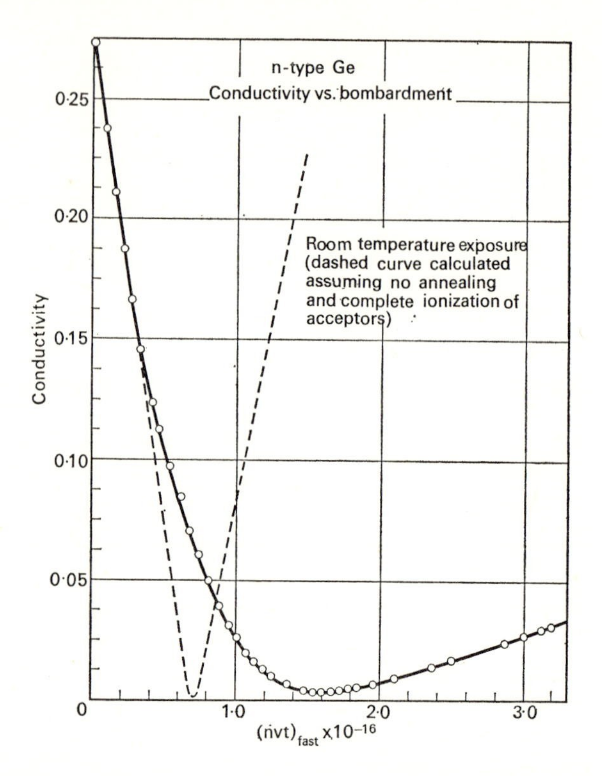

FIG. 1. Conductivity of *N*-type Ge vs. integrated fast neutron flux bombarded at room temperature. The dashed line indicates the expected behavior in the absence of annealing, variation in mobility, and complete ionization of introduced acceptors.

The results of these calculations have been reported previously for a number of *N*-type samples.[14] The average value of *K*, revised slightly in the light of additional data, is 3·2. Assuming a scattering cross-section of 1×10^{-24} cm^2 one finds the number of traps produced per scattered neutron to be 81. This value is in reasonable agreement with the calculations of G. E. Evans[15] from particle interaction theory who finds that 135 displaced atoms should be produced per scattered neutron using the energy distribution in the Oak Ridge Reactor. This agreement, however, depends on the choice of the scattering cross-section. These measurements seem to indicate that K is independent of temperature and initial impurity concentration.

The dashed curve in Fig. 1 was calculated on the assumptions that (1) all acceptors introduced by bombardment lie on the top of the filled band thus being equally effective in removing electrons or producing holes, (2) no annealing takes place, and (3) the mobility of the carriers is unaffected by bombardment. It is interesting to note that this has the same form as a simple conductiometric titration curve. The minimum value was calculated from Eq. (2). The early observed departure from linearity is readily explained on the basis of the aforementioned redistribution of trapped electrons seeking lower energy traps as these become available. Also any annealing of lattice disorder at room temperature, which is known to be appreciable from various indications to be discussed below, will cause the rate of electron removal to decrease with bombardment.

The minimum value of the conductivity of a homogeneous semiconductor in thermal equilibrium may be calculated directly from Eq. (4) by using the appropriate values of μ and κ. In a previous publication[16] it was pointed out that the observed value of σ_{min} was higher by a factor of 2 to 3 than the calculated value. This discrepancy was attributed to slight inhomogeneities in the impurity distribution throughout the sample. It should be stated, however, that the same effect could be caused by the photoelectric production of carriers by high intensity β- and γ-radiation. Experiments are under way to determine the influence of high intensity ionizing radiation on the electrical conductivity of Ge. As yet we have been able to determine no observable effect due to 1·3-MeV radiation even in high resistivity Ge ($\sim$ 5 ohm cm) though samples containing a rectifying potential barrier show large conductivity increases at liquid nitrogen temperature.

After the minimum is passed the conductivity increases at a rate much lower than the initial rate of decrease. The reduction in rate is due to two causes: (1) annealing of lattice defects and (2) only those acceptors which are ionized are effective in increasing the hole concentration. In this connection it is instructive to examine the ratio of slopes of the two legs of the conductivity vs. bombardment curve. In Table 1 the ratios of the initial slopes to those after conversion to *P*-type for a number of *N*-type samples bombarded at room temperature are listed, together with the initial electron concentration n_e^0 which, because of exhaustion,[17] is essentially equal to the original donor concentration. The slope ratio for $n_e^0 < 10^{13}$ (*P*-type Ge) was obtained from the mean value of the original *N*-type slope and that of *P*-type material at room temperature.[18] From these data it is evident that the ratio of slopes increases with increasing initial electron concentration. Since it has been shown previously[14] that the initial rate of change of conductivity for *N*-type Ge under bombardment at temperatures in the exhaustion range is constant, the variation of the ratio is due to the dependence of the slope after conversion to *P*-type on the initial electron concentration of the *N*-type material.

The dependence of the conductivity vs. bombardment slope

TABLE 1. THE RATIOS OF THE INITIAL SLOPE TO THAT AFTER THE CONDUCTIVITY MINIMUM OBTAINED FROM THE ROOM TEMPERATURE CONDUCTIVITY VS. BOMBARDMENT CURVES OF *N*-TYPE Ge. ALSO LISTED IS THE ORIGINAL ELECTRON CONCENTRATION n_e^0. THE *P*-TYPE VALUE ABOVE IS THE RATIO OF THE SLOPE OF AN *N*-TYPE SAMPLE WHICH IS CONSTANT FROM SAMPLE TO SAMPLE TO THAT OF A *P*-TYPE SAMPLE AT 30°C

Sample	Slope ratio	n_e^0
1	7·2	$4·2 \times 10^{14}$
2	8·0	$5·1 \times 10^{14}$
3	10·0	$1·3 \times 10^{15}$
4	12·9	$2·5 \times 10^{15}$
5	21·0	$1·0 \times 10^{16}$
6	21·4	$2·8 \times 10^{16}$
7	24·3	$5·5 \times 10^{16}$
P-type value	~ 6·5	$< 10^{13}$

after conversion may be caused by either the variation of disorder-annealing rate with the amount of disorder at the minimum, of which n_e^0 is an index, or the variation of the number of ionized acceptors introduced at the minimum which are occupied by electrons from the conduction band and donor levels. This latter effect is analogous to the common ion effect of solution chemistry. There is not sufficient evidence at the present time to make a clear choice. Both, quite probably, are in some manner responsible for the dependence of the slope ratio on the initial electron concentration of the *N*-type material. Examination of the data in Table 1 indicates that apparently ln n_e^0 is approximately proportional to the slope ratio. An elementary examination of the two explanations offered above, however, indicates that in both cases one would expect an approximately linear dependence between the ratio and n_e^0. It is encouraging to note that the ratio approaches the value 6·5 obtained for originally *P*-type material. A more elaborate treatment on the basis of more accurate data is necessary to explain this behavior quantitatively.

In order to minimize annealing during bombardment several *N*-type Ge samples were exposed at dry-ice temperature and the conductivity measured during bombardment. The conductivity vs. bombardment curve for one such sample is shown in Fig. 2. The behavior of the conductivity is qualitatively the same as at room temperature. The initial rate of removal of electrons is the same as that measured at higher temperatures within experimental error. The conductivity minimum occurs at a lower value than that observed at room temperature which is expected from the temperature dependence of κ in Eq. (4), though the observed value is higher by two orders of magnitude than the calculated value (3×10^{-3} ohm^{-1} cm^{-1} observed compared to 8×10^{-6} ohm^{-1} cm^{-1} calculated). The ratio of the initial rate of conductivity change to that after conversion is 12·5 and the initial donor concentration, as determined from the room temperature Hall constant, is $\sim 4 \times 10^{14}$. Comparison of these values with those of Table 1 shows that, for samples of comparable impurity concentration (samples 1 and 2), the rate of increase of carrier

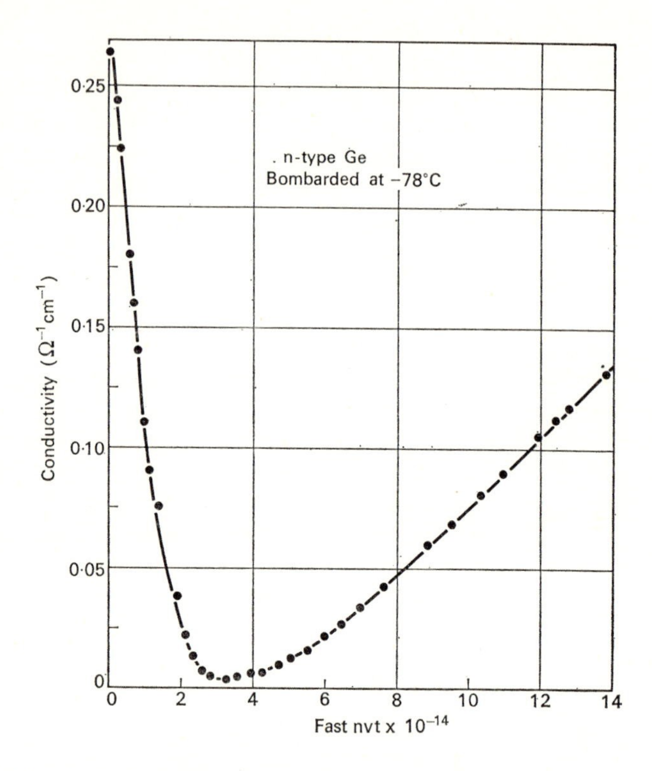

FIG. 2. Conductivity of *N*-type Ge vs. integrated fast neutron flux bombarded at dry ice temperature.

concentration is larger for higher temperatures. The temperature dependence of the rate of increase of holes will be considered further in connection with *P*-type Ge bombardment.

IV. THE EFFECT OF FAST NEUTRONS ON *P*-TYPE Ge

A typical conductivity vs. bombardment curve for *P*-type Ge at room temperature is shown in Fig. 3. The conductivity increases monotonically with bombardment and the initial slope is linear.

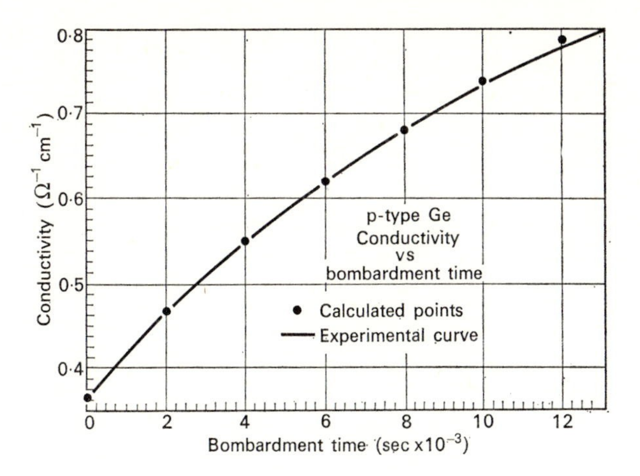

FIG. 3. Conductivity of *P*-type Ge vs. integrated fast neutron flux bombarded at room temperature. Dots represent the points calculated from Eq. (6).

As in the case of *N*-type Ge the number of carriers introduced per incident fast neutron may be calculated from the initial slope and the initial mobility under the same assumptions used in the *N*-type analysis. These calculations have also been previously reported,[16] and are reproduced with additional data in Table 2. The value at

TABLE 2. THE RATE OF INCREASE IN HOLE CONCENTRATION PER INCIDENT NEUTRON FOR *P*-TYPE MATERIAL AT VARIOUS TEMPERATURES. ALSO LISTED IS THE ORIGINAL HOLE CONCENTRATION AT 27°C

Sample (*P*-type Ge)	Temperature of exposure	Increase in hole concentration per incident neutron	Original hole concentration at 27°C (cm^{-3})
1	30°C	0·77	$1·7 \times 10^{15}$
2	20°C	0·70	$4·2 \times 10^{14}$
3	0°C	0·61	$2·5 \times 10^{14}$
4	− 78°C	0·20	$4·6 \times 10^{14}$
5	− 78°C	0·48	$4·9 \times 10^{13}$

30°C is about 0·8 hole per incident fast neutron. Thus only about one out of the four traps produced by bombardment is effective in increasing the hole concentration at this temperature. The high temperature rates (0°C to 30°C) appear to be essentially independent of initial carrier concentration but are evidently temperature dependent. It should be noted, however, that dependence on initial carrier concentration could be masked by a larger effect of temperature.

Further examination of the *P*-type Ge bombardment curve shows that the slope falls off with increasing bombardment, indicating that the conductivity tends toward saturation. This may be explained, at least in part, on the basis of annealing of lattice defects and is borne out by the decay of the conductivity increase during pile shut down. In fact, by assuming only one annealing process with a single activation energy, the above curve may be fitted rather well by the familiar first-order build-up equation

$$\sigma = \sigma_0 + (A/k')(1 - e^{-k't}), \tag{9}$$

where k' is the rate constant and A is a constant determined by the initial slope. The calculated points for $k' = 6\cdot2 \times 10^{-5}$ sec^{-1} are shown in Fig. 3. This result, however, is not very definitive since the value of k' depends on A. Because of pile power and temperature changes during start-up, the value of A is not sufficiently accurate for a refined determination of the rate constant. Attempts were made to calculate the curve using a simple second-order build-up equation

$$\sigma = \sigma_0 + (A'/k'^{\frac{1}{2}})\tanh[(A''k')^{\frac{1}{2}}t], \tag{10}$$

but no acceptable fit could be obtained. Similar analyses have been carried out by Brattain and Pearson[3] on Ge bombarded with polonium alpha-particles. Both build-up and decay curves were fitted with first-order equations and the rate constant thus obtained was $k = 1\cdot5 \times 10^{-5}$ sec^{-1}. These authors found that about 75% of the carriers introduced by bombardment was annealable at room temperature.

There is evidence that the annealing mechanism is much too complicated to be explained on such a simple basis as is discussed above. Preliminary experiments on the rate of annealing of Ge exposed at dry ice temperatures as well as examination of heat treatment data indicate that, quite probably, the annealing of lattice disorder involves multiple rate processes of different activation energies. There is even the possibility that there is a continuous spectrum of activation energies. A program intended to investigate this problem is now under way.

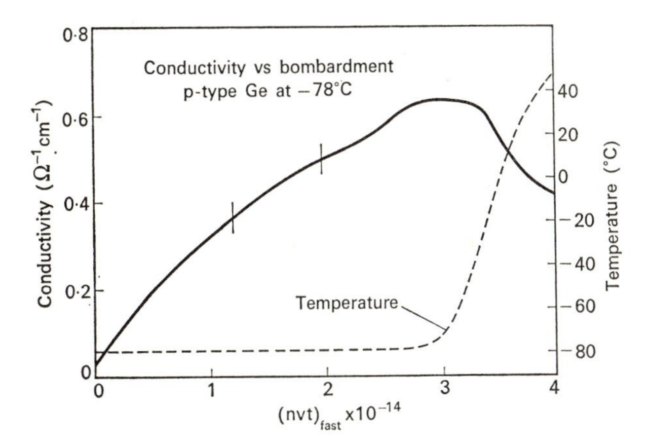

FIG. 4. Conductivity of *P*-type Ge vs. integrated fast neutron flux bombarded at dry ice temperature. The dashed line represents temperature. The vertical lines on the curve represent the range of coincidence with the curve of another sample (see text).

Two *P*-type samples were bombarded at dry ice temperature in order to decrease the effect of annealing. The conductivity vs. bombardment curve of the higher resistivity sample is shown in Fig. 4, and the pertinent data for the two samples are summarized in Table 2. The initial slopes for these samples indicate quite definitely that they are dependent on the initial carrier concentrations. The slope, in fact, seems to be dependent on the hole concentration at any point during the bombardment. This is borne out by the observation that, if one adjusts the $(nvt)_{fast}$ scale of the low resistivity sample so that its initial conductivity falls on the curve for the high resistivity sample (shown in Fig. 4), the curves coincide. This is indicated in Fig. 4 by two vertical marks on the curve between which the curves are coincident. Departure above the upper mark is due to temperature variation which sets in at this point.

During the low temperature bombardment discussed above, the conductivity was followed until all of the dry ice sublimed and the temperature rose toward pile ambient. The interesting behavior shown in Fig. 4 was observed. When the temperature (indicated by a dashed line) begins to rise the curve goes through an inflection and the rate of change of conductivity with bombardment begins to increase. The conductivity then reaches a maximum value with further temperature increase and then decreases. This behavior can be readily explained on the basis of simple semiconductor theory. The accelerated increase in conductivity when the temperature begins to rise is due to an increase in hole concentration. The decrease of conductivity at higher temperatures is caused by the decrease in lattice mobility with increasing temperature, which becomes predominant over the effect of increasing carrier concentration. It is also quite possible that an increased annealing rate caused by the temperature rise could anneal out a fraction of accumulated disorder at the high temperature, thus further decreasing the conductivity. This latter point, however, cannot be substantiated on the basis of existing data.

Perhaps the most interesting information to be obtained from the low temperature bombardment experiments is that even for very low annealing rates there is appreciable curvature in the *P*-type Ge conductivity vs. bombardment curve. This curvature is presumably due to the nonlinear dependence of the carrier concentration on acceptor concentration. In view of this behavior the validity of such a simple rate analysis of the build-up behavior of the conductivity during bombardment of *P*-type Ge is even less certain.

V. The Energy Distribution of Bombardment Introduced Acceptors

In order to study the energy distribution of acceptors produced by bombardment, measurements of the Hall coefficient R as a function of temperature on bombarded material are necessary. Because of the appreciable rate of annealing at moderate and high temperatures and because of the long "radioactivity cooling" period required before the samples can be handled, little can be learned concerning the energy distribution in this temperature range except for those acceptors which anneal at higher temperatures.

Figure 5 gives the log R vs. $1/T$ curve for a sample which had been stored at room temperature for several months after a long bombardment. Curve I was taken before any heat treatment, curve II after being heated to 400°C, curve III after 2 additional hours at 400°C and curve IV after 24 hours at 450°C. Further heat treatment produced no detectable change. Since a single ionization energy[19] in the impurity range would yield a constant slope, the curvature of curve I obviously indicates a wide distribution of ionization energies. It is interesting to note that nearly all of the effect of lattice disorder can be removed by merely heating to 400°C.

Additional information may be gained by exposing a sample at dry ice temperature and storing it in dry ice during the "cooling" period. By keeping the sample at this temperature and by first taking measurements at the low temperature, annealing of lattice disorder may be minimized. The results of two such exposures on the same sample are shown in Fig. 6. Hall coefficient curves are shown after various treatments. The sample was originally a high resistivity N-type Ge plate. The conductivity was followed during the bombardment and the sample was removed from the pile as near the conductivity minimum as possible in order to trap most of the conduction electrons. Curve I is the initial measurement before the first exposure. Unfortunately one of the Hall probes was detached during bombardment and had to be resoldered.

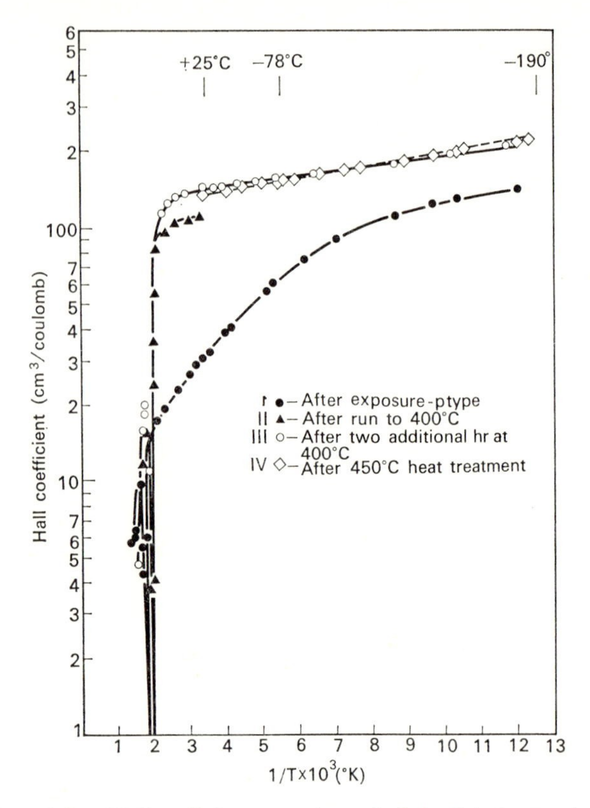

FIG. 5. Log Hall coefficient vs. reciprocal of absolute temperature of *P*-type Ge after long bombardment showing effect of various heat treatments.

This caused an uncertain amount of annealing and, therefore, curve II, taken after exposure, is not entirely representative. Curve III was taken after the sample was heated to 150°C and curve IV was taken after a full anneal.

After the second exposure the Hall coefficient had a positive sign indicating that the material had been converted to *P*-type [see Eq. (5)]. Because of rectification due to inhomogeneities in

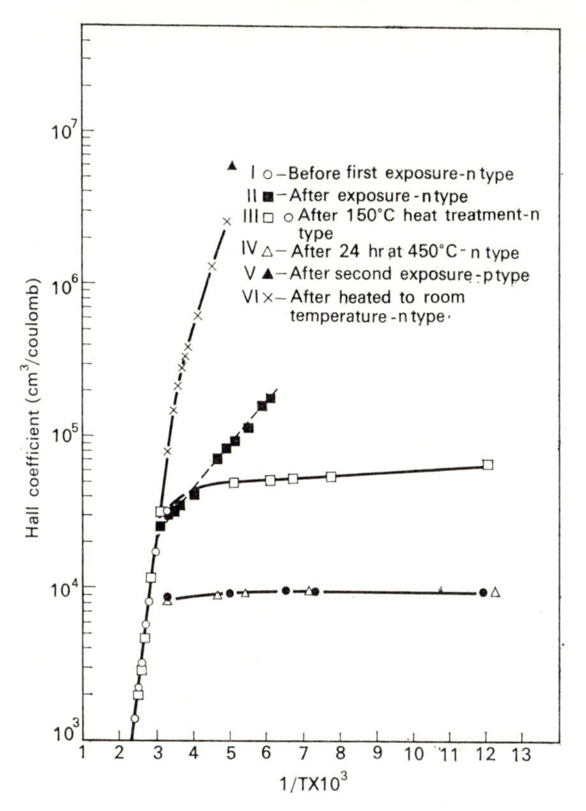

FIG. 6. Log Hall coefficient vs. reciprocal of absolute temperature of *N*-type Ge exposed to minimum conductivity at dry ice temperature after various heat treatments.

impurity distribution which, though unimportant in the original, produce rectifying *P–N* potential barriers[20] at the conductivity minimum, reliable Hall coefficient measurements could not be made. The sample was briefly warmed to room temperature and recooled. By this action sufficient annealing took place to cause the sample to revert to *N*-type and to reduce the low temperature rectification. Curve VI was taken on this material.

Apparent ionization energies calculated from the low temperature slopes of curves II and VI are respectively 0·14 and 0·31 eV. These ionization energies are only apparent since they represent the effective energy of electrons in a distribution of traps. Presumably curvature like that in curve II, Fig. 5 would appear if the measurements could have been extended to lower temperature. Noting that the sample in curve II has had more heat treatment than in curve VI, this seems to indicate that annealing of lattice disorder in *N*-type Ge near the minimum tends to decrease the apparent activation energy. If one assumes that these bombardment produced acceptors are distributed in energy, uniform annealing will have the effect of promoting electrons to the more shallow traps. As the annealing progresses the concentration of electrons associated with the original donors increases and the ionization energy tends to revert to that of the original sample even though there is a considerable concentration of acceptors yet remaining. This is the case for curve III.

VI. Summary and Conclusions

Fast neutron bombardment of Ge introduces into the forbidden energy band a distribution of electron traps. These traps are associated with lattice defects, presumably lattice vacancies, caused by the bombardment. Hole traps may also be produced, but, if these are present in Ge, they lie so near the top of the filled band that they have no observable effect on the electron concentration. The introduced traps remove electrons from *N*-type Ge initially at a uniform rate (3·2 electrons removed per incident fast neutron of the energy spectrum considered here). As intrinsic behavior is approached the trapped electrons are redistributed moving to traps of lower energy, until a sufficient concentration of low energy traps is introduced to contain all of the impurity electrons. Further bombardment causes a transition to *P*-type material and the low-lying traps behave as acceptors. Experimental evidence indicates that at room temperature one out of about four of the introduced traps is sufficiently deep lying to produce positive carriers. The conductivity vs. bombardment curve resembles a conductiometric titration curve between a strong and weak electrolyte. Bombardment of *P*-type Ge causes a monotonic increase in conductivity with concave downward curvature indicating an approach to saturation, provided the Fermi level does not lie so deep as to be insensitive to additional acceptors of finite ionization energy.

Bombardment-produced disorder, like disorder produced by quenching, can be annealed out by proper heat treatment. All of the effects on electrical properties of Ge can be removed by a vacuum anneal at 450°C. Considerable annealing at room temperature is also observed. A simplified analysis of the *P*-type bombardment curve indicates that recombination of defects takes place through a first-order process, but other evidence seems to indicate that the annealing kinetics are too complicated to be fruitfully treated by such a naive approach.

The purpose of the investigations reported here is twofold: (1) the use of the fast neutron bombardment technique to investigate the effect of lattice disorder on the energy level scheme of Ge, and (2) the use of semiconductors whose properties are sensitive to small amounts of disorder to investigate the fundamental nature of radiation damage. The results of these experiments indicate that further thorough study of the energy distribution of electron traps produced by bombardment and the kinetics of annealing of the lattice disorder should be of considerable value in elucidating both aspects of the problem. Experiments designed to accomplish these ends are now under way.

Evidence for Production of Hole Traps in Germanium by Fast Neutron Bombardment

In Si[21] both acceptors or electron traps and donors or hole traps, presumably associated with lattice vacancies and interstitials, respectively, are introduced near the middle of the forbidden band by nucleon bombardment. Consequently, the Fermi level ζ of *N*-type Si is depressed and that of *P*-type Si is elevated by bombardment toward a limiting or saturation value ζ_{limit} causing a corresponding decrease in the conductivity of both *N*- and *P*-type Si. Heretofore, experiments on both *N*- and *P*-type Ge[22] indicate that bombardment causes an increase in *P*-type character with an accompanying depression of ζ toward the top of the filled band. This behavior may be interpreted as being the result of the introduction of acceptors only. However, in view of the close similarity of the electronic and crystal structure of Ge and Si, one would expect donors to be introduced into Ge as well.

James and Lehman[23] have investigated the situation in which both acceptors and hole traps (donors) are introduced in equal numbers below the middle of the filled band and find that ζ_{limit} lies half way between the two introduced levels. Thus the ζ of originally *N*-type or high to moderate resistivity *P*-type Ge will be depressed toward ζ_{limit} by bombardment while low resistivity *P*-type Ge with a ζ value below ζ_{limit} will show an elevation of ζ corresponding to a decrease in conductivity.

TABLE 1. BOMBARDMENT DATA FOR LOW RESISTIVITY *P*-TYPE Ge

Sample	n_h^0	$dn_h/d(nvt)_{fast}$	T_e
1	$7{\cdot}0 \times 10^{17}$ cm^{-3}	$-5{\cdot}0$	32°C
2	$1{\cdot}01 \times 10^{19}$	$-5{\cdot}0$	37
3	$1{\cdot}3 \times 10^{19}$	$-2{\cdot}2$	48
4	$1{\cdot}5 \times 10^{19}$	$-2{\cdot}9$	48

The conductivity of 5 low resistivity *P*-type Ge single crystals[24] ($n_h > 7 \times 10^{17}$ cm^{-3}) was followed during fast neutron bombardment in the Oak Ridge pile. The initial rate of change in carrier concentration per incident neutron $dn_h/d(nvt)_{fast}$, the original hole concentration n_h^0, and the exposure temperature for these samples are listed in Table 1. All samples showed a decrease in conductivity with bombardment and a gradual approach to a limiting value as predicted by the model.

The limiting value of the hole concentration, above which a decrease in hole concentration with bombardment is expected, is given by

$$(n_h)_{\text{limit}}=2(2\pi m_h{}^*kT/h^2)^{\frac{3}{2}}\exp(-\zeta_{\text{limit}}/kT], \qquad (1)$$

where ζ_{limit} is measured from the top of the filled band. Consequently, it should be possible to choose Ge samples with appropriate hole concentrations which on bombardment would show a decrease in carrier concentration at low temperatures and an increase at high temperatures within a convenient temperature range. Several samples with hole concentrations of $\sim 10^{16}$ cm^{-3} were bombarded at $-$ 78°C, and the conductivity was followed at that and at ambient temperature (55°C) after the dry ice was sublimed. A typical conductivity vs. integrated fast neutron flux $(nvt)_{\text{fast}}$ curve is shown in Fig. 1. The values of $dn_h/d(nvt)_{\text{fast}}$ for both temperatures and the original hole concentrations for these samples are listed in Table 2. In every case the slope is negative at $-$ 78°C and positive at 55°C in agreement with theory. Thus the James–Lehman model is qualitatively verified by these experiments.

TABLE 2. BOMBARDMENT DATA FOR *P*-TYPE SAMPLES BOMBARDED SUCCESSIVELY AT $-$ 78°C AND 55°C

Sample	$n_h{}^0$ ($-$ 78°C)	$dn_h/d(nvt)_{\text{fast}}$ at $-$ 78°C	at 55°C
1	$4 \cdot 01 \times 10^{16}$	$-0 \cdot 99$	$0 \cdot 20$
2	$4 \cdot 69 \times 10^{16}$	$-1 \cdot 12$	$0 \cdot 396$
3	$4 \cdot 86 \times 10^{16}$	$-1 \cdot 03$	+
4	$7 \cdot 23 \times 10^{16}$	$-1 \cdot 10$	+

An idea of the magnitude of ζ_{limit} for Ge may be obtained from Eq. (1) and the experimental data. The values of $n_h{}^0$ for sample 1 of Table 1 and sample 4 of Table 2 may be used to bracket $(n_h)_{\text{limit}}$. At 55°C $7 \cdot 23 \times 10^{16}$ cm^{-3} $< (n_n)_{\text{limit}} < 7 \cdot 02 \times 10^{17}$ cm^{-3}, and from Eq. (1) it can be shown that $0 \cdot 168$ eV $> \zeta_{\text{limit}} > 0 \cdot 105$ eV. Further experiments designed to determine ζ_{limit} more exactly and to test the model more rigorously are under way.

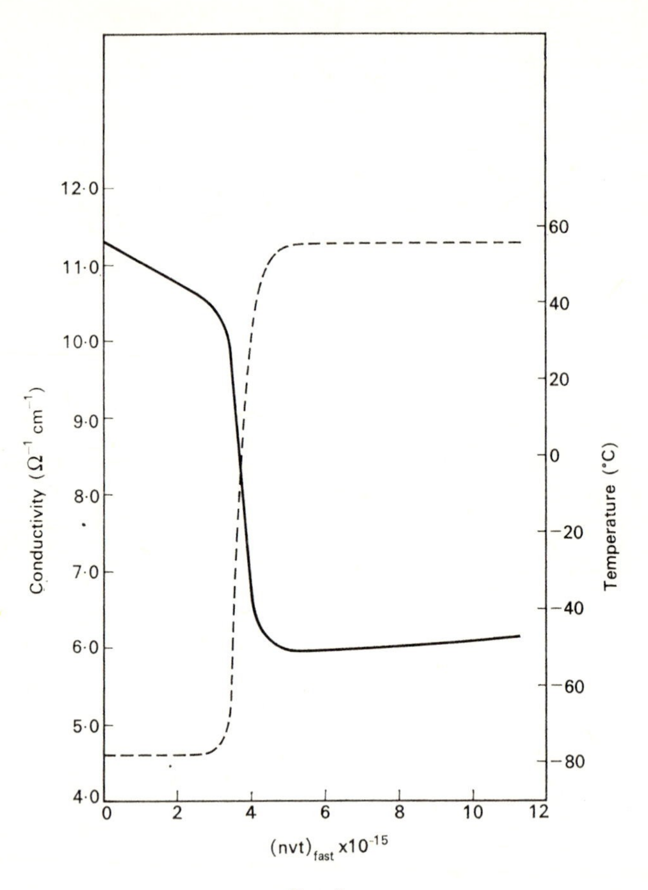

FIG. 1

NOTES

[1] Lark-Horovitz, Bleuler, Davis, and Tendam, *Phys. Rev.* **73,** 1256 (1948). See also Purdue Progress Reports to Signal Corps, Nov. 1947 to present (unpublished).

[2] Davis, Johnson, Lark-Horovitz, and Siegel, *Phys. Rev.* **74,** 1255 (1948); W. E. Johnson and K. Lark-Horovitz, *Phys. Rev.* **76,** 442 (1949). See also Davis, Johnson, Lark-Horovitz, and Siegel, AECD Report 2054, June 1948 (unpublished).

[3] N. H. Brattain and G. L. Pearson, *Phys. Rev.* **80,** 846 (1950).

[4] Cleland, Lark-Horovitz, and Pigg, *Phys. Rev.* **78,** 814 (1950).

[4a] In the case of bombardment of Si the effect of both electron and hole traps is equally important in contrast with the case of Ge. The conductivity of both *N*-type and *P*-type silicon decreases with bombardment and each appears to approach a limiting value. A report concerning the behavior of reactor irradiated silicon will be forthcoming in the near future.

[5] K. Lark-Horovitz, NDRC Report 14–585, covering the period from March 1942 to November 1945, p. 24 (unpublished). For more recent developments see W. E. Taylor, doctoral thesis, published as a special report to Signal Corps, Purdue University, June 1950, and *J. Metal Tech.* (to be published).

[6] F. Seitz, *Disc. Faraday Soc.* No. **5,** 271 (1949).

[7] V. Vand, *Proc. Phys. Soc.* (London) **55,** 222 (1943).

[7a] The probability of the existence of hole traps or donors has been noted previously. These are not considered in the above discussion, since in Ge they presumably lie near the top of the filled band and, therefore, are not effective in releasing electrons or trapping holes. However, if these hole traps lie above acceptors, their net effect will be to increase the ionization energy of available holes. For an account of the effect of hole traps on the energy distribution in semiconductors see Note 8.

[8] K. Lark-Horovitz, Appendix II (Lehman), International Conference on Semiconductors, Reading, 1950; see also G. Lehman, *Phys. Rev.* **81,** 321 (1951).

[9] E. Conwell and V. F. Weisskopf, *Phys. Rev.* **69,** 258 (1946); **77,** 388 (1950).

[10] V. A. Johnson and K. Lark-Horovitz, *Phys. Rev.* **69,** 258 (1964).

[11] G. L. Pearson and J. Bardeen, *Phys. Rev.* **75,** 865 (1949); C. Erginsoy, *Phys. Rev.* **78,** 1013 (1950).

[12] V. A. Johnson and H. Y. Fan, *Phys. Rev.* **79,** 899 (1950).

[13] K. Lark-Horovitz, Appendix I (Johnson and Lark-Horovitz), International Conference on Semiconductors, Reading, 1950.

[14] J. H. Crawford, Jr., and K. Lark-Horovitz, *Phys. Rev.* **78,** 815 (1950).

[15] G. E. Evans (to be published).

[16] J. H. Crawford, Jr. and K. Lark-Horovitz, *Phys. Rev.* **79,** 889 (1950).

[17] The exhaustion region is that region in temperature at which essentially all of the impurity atoms are ionized.

[18] Since it has been shown that the initial *N*-type conductivity slope is constant from sample to sample except for mobility variations [Eq. (8)] the ratio of this value to the room temperature value for a *P*-type sample should be an index of the slope ratio as defined above for $n_e^0 < 10^{13}$.

[19] The term "ionization energy" is preferred instead of the more frequently used "activation energy" since the latter has special application in rate process theory.

[20] Orman, Fan, Goldsmith, and Lark-Horovitz, *Phys. Rev.* **78,** 645 (1950).

[21] W. E. Johnson and K. Lark-Horovitz, *Phys. Rev.* **76,** 442 (1949).

[22] Cleland, Crawford, Lark-Horovitz, Pigg, and Young, *Phys. Rev.* **83,** 312 (1951).

[23] K. Lark-Horovitz, Appendix II (Lehman and James), International Conference on Semiconductors, Reading, 1950; see also G. W. Lehman, *Phys. Rev.* **81,** 321 (1951).

[24] The samples used in these investigations were kindly prepared and furnished by Miss Louise Roth, Purdue University.

18

The Effect of Neutron Bombardment on the Low Temperature Atomic Heat of Silicon†‡

EXPERIMENTS at Purdue with 10 MeV deuterons and at Oak Ridge with fast neutrons have indicated that irradiation produces both electron and hole traps in silicon, since both *N*- and *P*-type materials become very poorly conducting (resistivities as high as 250,000 ohm-cm have been reached in this way). Thus a result of the lattice disorder (vacant sites and interstitial atoms) produced by elastic collisions is the removal of free carriers.[1,2] Since annealing at 450°C restores the initial conductivity, it appears to remove the electron and hole traps and to release the trapped carriers; hence, any electronic (or hole) contribution to the atomic heat at very low temperatures should be decreased by bombardment and then restored by annealing at elevated temperatures.

The lattice contribution to the atomic heat at very low temperatures should also be sensitive to bombardment because of the dependence of the thermal vibrations on the interatomic forces, which in turn are affected by the disordering of the lattice.[3-5] This has been known for some time from the investigation of the specific heat of the so-called metamict crystals,[6] in which the bombardment is produced by radioactive inclusions in the lattice.

In order to see whether such changes could be observed experimentally, the atomic heat of a silicon ingot§ was measured before and after exposure to about $5 \times 10^{18}/cm^3$ neutrons in the nuclear

† *Science* **116,** 630 (1952), with P. H. Keesom and N. Pearlman.

‡ We would like to thank D. Finlayson for his measurements of the Hall constant, and L. Roth for the probe measurements and heat treatments.

§ The ingot weighed about 260 g, and was kindly supplied by G. L. Pearson, of the Bell Telephone Laboratories. Its impurity concentration was not known.

reactor at Oak Ridge. The sample was then annealed at several temperatures up to 780°C (Table 1), being held at each temperature for the time indicated, and then cooled at about 10°C/hr. Its heat capacity below 5°K was measured after each heat treatment.

TABLE 1. LATTICE AND CARRIER CONTRIBUTIONS TO ATOMIC HEAT OF SILICON AS A FUNCTION OF BOMBARDMENT AND HEAT TREATMENT

	Run	N	θ (°K)	γ-J/mole degree^2
1	Si II—before bombardment	18	645	$13 \cdot 03 \times 10^{-6}$
2	,, ,, ,, ,,	33	659	22·51
3	,, ,, ,, ,,	35	661	22·51
	Runs 1, 2, and 3 combined	86	658	21·02
4	Si IIB—after bombardment	18	633	1·11
5	,, ,, ,, ,,	20	640	1·96
	Runs 4 and 5 combined	38	637	1·55
6	Si IIB (A1)—anneal at 135°C (24 hr)	15	628	− 0·91
7	Si IIB (A2)—anneal at 253°C (24 hr)	17	632	0·43
8	Si IIB (A3)—anneal at 472°C (24 hr)	16	628	− 2·08
9	Si IIB (A4)—anneal at 455°C (48 hr)	13	637	1·74
	Runs 6, 7, 8, and 9 combined	61	632	− 0·03
10	Si IIB (A5)—anneal at 780°C (24 hr)	14	642	13·69

The experimental method was similar to that of Nernst and Eucken[7] and has been described in detail elsewhere.[8] At very low temperatures, the contribution of the lattice to the atomic heat will be given by[9]

$$1944(T/\theta)^3 = \alpha T^3 \text{ J/mole degree,} \tag{1}$$

where θ is the Debye temperature. The contribution of the carriers can be written[10]

$$1{\cdot}62 \times 10^{-12} V n^{1/3} \mu T = \gamma T \text{ J/mole degree,} \tag{2}$$

where V is the atomic volume, n is the carrier concentration per cm^3, and μ is the ratio of carrier mass to electron mass. Other evidence indicates that this ratio is about unity for carriers in silicon;[11] hence, the total atomic heat should have the form

$$C_v = \alpha T^3 + \gamma T \quad (3)$$

at very low temperatures. We verified this form of temperature-dependence for our data by plotting C_v/T vs. T^2 and observed that the points scattered about a straight line for each run. The slope α and intercept γ were calculated by least squares, and θ (calculated from α using Eq. (1)) and γ are listed in Table 1. The number of heat capacity measurements made during each run is given in the third column of the table.

The electronic term became very small after bombardment and remained small after the first four anneals. There would appear to be no significant change in the electronic term after any of these treatments, whereas the last anneal restores the original linear term, within the accuracy of our measurements. We attempted to determine n, the carrier concentration, independently from measurements of the Hall constant, but the agreement found with n determined from Eq. (2) (about 10^{18} cm^{-3} before bombardment) is inconclusive, since probe measurements of the large ingot showed it to be very inhomogeneous, with both N- and P-type regions of high and low conductivity.

The effect on the Debye temperature is not as clear, although it also appears to decrease after bombardment and to rise after the last heat treatment. The initial decrease would imply that the decrease of elastic constants attributable to vacancies predominates over the increase due to interstitial atoms. Dienes' calculations for copper and sodium[3,5] show that this situation would be favored in an open lattice, such as the diamond structure in which silicon crystallizes.

Our results thus lend further support to the idea that lattice disorder produced by neutron bombardment results in carrier traps in silicon which can be healed out by suitable heat treatment. It also confirms the division of the total atomic heat into two parts, as indicated by Eq. (3), and the identification of the linear term with the contribution of the carriers.

REFERENCES

1. LARK-HOROVITZ, K. In H. K. Henisch (Ed.), *Semi-Conducting Materials*, New York: Academic Press, pp. 41–70 (1951).
2. JAMES, H. M., and LARK-HOROVITZ, K. *Z. physik. Chem.* **198,** 107 (1951).
3. DIENES, G. J. *Phys. Rev.* **86,** 228 (1952).
4. NABARRO, F. R. N. *Ibid.* **87,** 665 (1952).
5. DIENES, G. J. *Ibid.* **87,** 666 (1952).
6. FAESSLER, A. *Z. Krist.* **104,** 81 (1942).
7. EUCKEN, A. *Handbuch der Experimentalphysik*, Vol. VIII—1. Leipzig: Akademische Verlag, Ch. 3 (1929).
8. PEARLMAN, N., and KEESOM, P. H. *Phys. Rev.* **88** 398 (1952).
9. DEBYE, P. *Ann. Physik* **39,** 789 (1912).
10. SOMMERFELD, A. *Z. Physik* **47,** 1 (1928).
11. PEARSON, G. L., and BARDEEN, J. *Phys. Rev.* **75,** 865 (1949).

Note added in proof. Another more homogeneous B.T.L. Si sample (SiV) has been investigated, and the results obtained agree with the results given above: $\theta_{orig} = 658°K$, $\theta_{bomb} = 636°K$; $\gamma_{orig} = 34{\cdot}6 \times 10^{-6}$, $\gamma_{bomb} = 9{\cdot}26 \times 10^{-6}$. Heat treatment again restores the electronic specific heat.

the carrier concentration by orders of magnitude even to the extent of changing the type of carriers, electrons, or holes. Therefore the electrical properties of these materials give a much more sensitive indication of the radiation effects (Lark-Horovitz, 1951). Furthermore, with small carrier concentrations semiconductors often show measurable transmission for light of sufficiently long wavelengths, usually in the infra-red region (Fan and Becker, 1951). Optical and photoconductive properties can also be used to study defects in the crystal as in the case of insulators (Becker, Fan and Lark-Horovitz, 1952; Fan, Kaiser, Klontz, Lark-Horovitz and Pepper, 1954). We shall discuss primarily the electrical properties as determined by resistivity and Hall coefficient measurements, based on some recent results obtained with germanium irradiated by α-particles, deuterons, and electrons. The experiments with 9·6 MeV deuteron irradiation were carried out using the Purdue cyclotron (Forster, Fan and Lark-Horovitz, 1952; Forster, 1952), and the 4·5 MeV electron irradiations with the Purdue linear accelerator (Pepper, Klontz, Lark-Horovitz and MacKay, 1954). Results for electrons with energies less than 1·8 MeV were obtained using the Van de Graaff generators at Notre Dame (Klontz and Lark-Horovitz, 1951a, b, 1952).

Qualitatively the effects observed are the same when irradiating with any of these charged particles, and also with fast neutrons: n-type germanium is converted to p-type. The conductivity of high resistivity p-type germanium increases, the conductivity of low resistivity p-type decreases. Heat-treatment at 450–500°C restores the original conductivity if only lattice defects were produced. We illustrate this behaviour by recent experiments of Becker and Lark-Horovitz on polonium alpha-particles irradiation of single crystal layers of germanium, a few microns thick, produced by alternately grinding and etching. The left half of Fig. 1 shows the irradiation curves (both conductivity and Hall constant) produced with a flux of about 0·4 millicurie falling on the sample. In this high resistivity n-type sample, at first the conductivity decreases, reaches a minimum in about 15 minutes, then increases again. There is a striking difference in slope on the

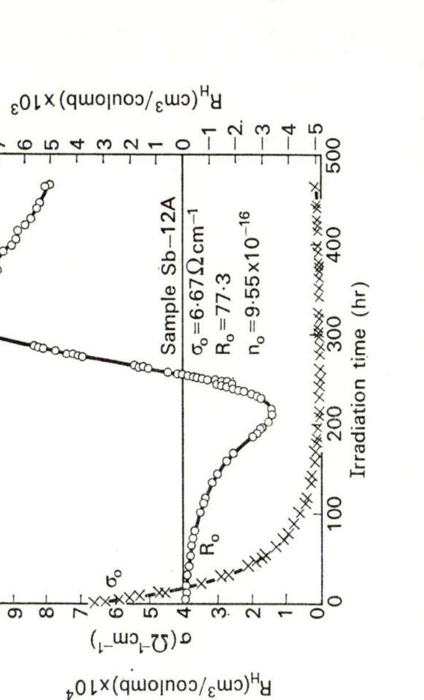

FIG. 1. Conductivity and Hall coefficient measurements on germanium semiconductors irradiated at room temperature with less than 5 MeV alpha-particles. The left half indicates the behaviour of a high resistivity sample, the right half the behaviour of a low resistivity sample (W. M. Becker and K. Lark-Horovitz).

negative and positive side. The Hall effect (negative) increases to a negative maximum in about 5 *minutes* and goes through zero after about twenty minutes.

The right half of the figure shows conductivity and Hall effect as a function of irradiation in a high conductivity germanium sample. The time necessary to reach the conductivity minimum is now much longer, but the efficiency of carrier removal per flux incident per unit area is far greater (Table 1).

TABLE 1. RATE OF REMOVAL OF CONDUCTION ELECTRONS AND CALCULATED DISPLACEMENTS OF ATOMS

Particle	Experimental					Calculated		
	E (MeV)	T (°K)	n_0 (cm^{-3})	$-dn/d\phi$ (cm^{-1})	$-dn_t/d\phi$	$\sigma(E)$ (barns)	$dN/d\phi$ (cm^{-1})	$dN_t/d\phi$
D	9·6	205	$4 \cdot 9 \times 10^{17}$		17	6350	910	17
	1·5	90	1×10^{16}	1·3		37	1·6	
e	4·5	205	1×10^{15}	3·7		65	4·9	
		90	1×10^{16}	8				
α	<5	R.T.	1×10^{17}		95			59
			9×10^{13}		22			

2. Correlation of Change in Carrier Concentration with Estimated Production of Defects

The various effects to be expected from the passage of high energy particles through germanium and silicon have been discussed by Lark-Horovitz (1951). Most of the energy is spent in exciting the electrons of the solid, which does not lead to changes of the crystal lattice in these materials. This is shown by the fact that whereas large changes in conductivity are observed during the irradiation of germanium by 0·34 MeV electrons, this change vanishes in a few microseconds after the irradiation, and no permanent effect is produced (Lark-Horovitz, 1951). With neutron irradiation a certain amount of transmutation is produced, changing the properties of the specimen irreversibly (Cleland, Lark-Horovitz and Pigg, 1950). This effect is negligible when irradiating with charged particles.

The main effect on the crystal lattice is displacement of atoms from regular lattice sites, produced by collision with the high energy particles.

Using electrons of different energies a threshold of 0·63 MeV was determined as necessary to produce the effect (Lark-Horovitz, 1951; Klontz and Lark-Horovitz, 1951a, b, 1952). Thus the recoil energy required for the displacement of a germanium atom was found to be $E_r = 30 \pm 1$ eV.

The number of displacements produced by a deuteron can be estimated in the following way. Using a value of E_r, the cross-section for displacement $\sigma(E)$ can be expressed as a function of deuteron energy E according to the Rutherford scattering formula. The range–energy relation for the stopping of deuterons in germanium has been determined by Heller and Tendam (1951):

$$R = 6{\cdot}33 \times 10^{-4} E^{1{\cdot}63} \quad \text{for} \quad 2\ \text{MeV} < E < 10\ \text{MeV}$$

where R is the range in cm and E is the deuteron energy in MeV. Thus the number of primary displacements is

$$N_p = \int_{R_1}^{R_0} \sigma(E)\, dR = 1{\cdot}03 \times 10^{-3} \int_{E_1}^{E_0} \sigma(E) E^{0{\cdot}63}\, dE$$

where E_0 and E_1 are the energies of incident and outgoing deuterons respectively. Since the recoil atoms have considerable energies they in turn may produce displacements. The total number of displacements resulting from a primary displacement can be calculated according to Seitz (1949).

For electron irradiation the number of primary displacements can be calculated by integrating Mott's expression (Mott, 1929) for the differential cross section for scattering between θ_0 and π where the angle θ_0 is determined by E_r. Using the energy E_r necessary to remove a germanium atom from a lattice site and the average energy $\bar{E}$ imparted to a germanium atom by collision with a 4·5 MeV electron, the displacements can be calculated from Seitz's theory.

The number of conduction electrons as we have seen in *n*-type germanium is reduced by irradiation. The correlation between the number of removed carriers, determined experimentally by the changes in resistivity and Hall coefficient, and the number of displacements calculated can be seen in Table 1. Some results on polonium α-particle irradiation at room temperature are also included. In the case of electron irradiation the thickness of the samples was much smaller than the particle range, and the data given refer to carrier concentration n and displacement concentration N per cm^3. For deuteron and α-particle irradiations the thickness of the samples was comparable to the particle range. Consequently the irradiation of the samples might not have been quite uniform through the thickness. These data refer to number of carriers n_t and number of displacements N_t in a sample of thickness t per unit area of the irradiated surface. To avoid complications that may arise in heavily irradiated samples, the rate of carrier removal is calculated from the initial slopes of (n_t, ϕ) curves. We see that there is a fair correlation between the measured carrier reduction and the calculated number of displacements. This agreement might lead one to the conclusion that each displacement produces an acceptor level which should be perhaps ascribed to the vacancy left behind. We shall see, however, that the situation is actually much more complicated.

3. The Dependence of the Irradiation Effect on Fermi-level

It should be pointed out that not only acceptors but also donors are introduced by irradiation. While the number of conduction electrons in *n*-type germanium is decreased by irradiation, the number of holes in *p*-type germanium is also reduced by irradiation if it is sufficiently large at the start. Figure 2 shows the conductivity curves for an *n*-type and *p*-type sample, irradiated by deuterons at 200°K (Forster, Fan and Lark-Horovitz, 1952;

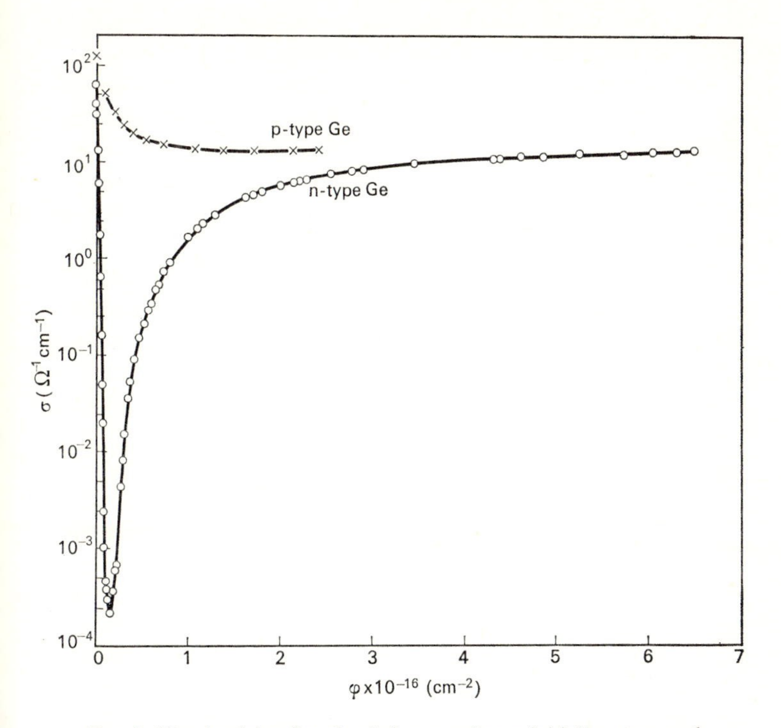

FIG. 2. (Conductivity–flux density) curves for an initially p-type and an initially n-type sample. Irradiation by 9·6 MeV deuterons at 200°K. (Measurements by J. H. Forster.)

Forster, 1952). We see that the conductivities in both samples decreased with irradiation, showing clearly that donors as well as acceptors were introduced. For n-type samples the conductivity reaches a minimum and then conduction changes to p-type, as shown by Hall coefficient measurement; further irradiation increases the hole concentration. With prolonged irradiation all samples tend to a constant hole concentration irrespective of original condition. This behaviour is also illustrated by the curves in Fig. 2: the same final hole concentration is reached through n-type and p-type bombardment. Therefore, using a sample of suitably chosen hole concentration no change of carrier concentration would be produced by irradiation at a given temperature.†

The electron occupation of the donors and acceptors associated with lattice defects depends upon the energy levels of these states relative to the Fermi-level ζ. Therefore the rate of change in carrier concentration produced by a given irradiation depends on the Fermi-level or the carrier concentration of the sample.

Table 2 gives the rate of change in carrier concentration for *n*-type and *p*-type samples of different initial carrier concentrations. The data were obtained from the initial slopes of the irradiation curves with fluxes less than 10^{15} cm^{-2}, in order to avoid complications possible after a large number of defects was produced. The large variation in the correlation between the number of calculated displacements and observed changes in carrier concentration in this table shows that such data can only be interpreted if the carrier concentration or the Fermi-level is known. Earlier experiments (Lark-Horovitz, Bleuler, Davis and Tendam, 1948) were carried out with high resistivity samples, to obtain effects in short bombardment times. For this reason the efficiency of carrier removal was far below expectation. In Table 1 we have chosen data obtained on samples with high carrier concentration or high Fermi-level, so as to approach the maximum rate of electron removal. The two different results for α-particle bombardment given there clearly show the effect of ζ.

The dependence of irradiation effects on the Fermi-level is determined by the statistics of electron distribution in thermal equilibrium and the condition of electrical neutrality which can be expressed as number of conduction electrons + number of ionized chemical acceptors + number of ionized irradiation introduced acceptors = number of conduction holes + number of

† At low temperatures (about 90°K) samples available ordinarily have a hole concentration too high to fulfil this condition, and one observes always hole removal in *p*-type samples. However, in gold-doped germanium with a high activation energy of acceptors, the hole concentration at about 90°K is so low that electron irradiation increases the hole concentration (Pepper, Klontz, Lark-Horovitz and MacKay, 1954).

TABLE 2. INITIAL RATES OF CARRIER REMOVAL BY IRRADIATION FOR SAMPLES WITH DIFFERENT CARRIER CONCENTRATIONS OR FERMI-LEVEL ζ

9·5 MeV Deuterons $T = 205°K$						9·5 MeV Deuterons $T = 90°K$		
n-type			p-type			p-type		
n_0 (cm^{-3})	$E_C - \zeta$ (eV)	$-dn/d\phi$ (cm^{-1})	p_0 (cm^{-3})	$\zeta - E_V$ (eV)	$-dp/d\phi$ (cm^{-1})	p_0 (cm^{-3})	$\zeta - E_V$ (eV)	$-dp/d\phi$ (cm^{-1})
$1·04 \times 10^{15}$	0·167	25·8	$4·8 \times 10^{15}$	0·137	20·8	$7·22 \times 10^{15}$	0·05	42
$1·58 \times 10^{16}$	0·116	228	$8·5 \times 10^{15}$	0·127	17·3	$1·22 \times 10^{18}$	0·01	260
$4·86 \times 10^{17}$	0·058	1130	$1·04 \times 10^{18}$	0·047	200			

E_C is the lowest energy in the conduction band, E_V is the highest energy in the valence band; n_0, p_0 are initial concentrations of carriers at the bombardment temperature; $E_C - \zeta$ or $\zeta - E_V$ measures the distance in eV for the ζ level from the conduction and the valence band; $-dn/d\phi$ and $-dp/d\phi$ are the number of electrons or holes removed per incoming deuteron as determined from the initial slope.

ionized chemical donors + number of ionized irradiation introduced donors.†

$$n - p = N_d - N_a + \sum_m \frac{D_m}{1 + 2 \exp (\zeta - E_m)/kT} - \sum_n \frac{A_n}{1 + 2 \exp (E_n - \zeta)/kT} \quad (1)$$

where ζ is the Fermi-level, n and p are electron and hole concentrations, respectively, D_m's and A_n's are the concentrations of various donor and acceptor states introduced by bombardment, N_d and N_a are the concentrations of impurity donors and acceptors. The summation of D terms gives the ionized donors introduced by irradiation and the summation of A terms gives the ionized acceptors. Considering the commonly used doping impurities, their donor and acceptor states are almost completely ionized in the temperature range used. The Fermi-level is uniquely related to n and p, which are determined experimentally. Equation (1) can be used, therefore, to deduce from experimental results information on the volume concentration and energy levels of localized states produced by irradiation.

4. Localized States in Irradiated Germanium

We may expect that vacancies in the crystal lattice give acceptor states, whereas the interstitial atoms give donor states. James and Lark-Horovitz (1951) have argued that each vacancy gives two acceptor states, one of which is close to the valence band, while each interstitial atom gives two donor states, one of which is close to the conduction band. Electron irradiation may be expected to produce simply vacancies and interstitials. However, if some of the interstitial atoms should come to rest near the vacancy left behind,

† The probability of occupation of a donor for electrons or an acceptor for holes is

$$f = [1 + \tfrac{1}{2} \exp \{\pm (E - \zeta/kT)\}]^{-1}.$$

An ionized donor or ionized acceptor corresponds to the probability of occupation $1 - f$ as given in (1).

the effect may not be that of isolated vacancies and interstitials. In the case of deuteron irradiation where a number of displacements can be produced by a primary recoil atom, clusters of vacancies and interstitials may be formed, which complicate the picture. Furthermore, on warming the sample different types of defects may anneal out at different rates and new arrangements of defects may be formed. Experimentally, the different energy levels and concentrations of associated localized states produced during the irradiations and their variation with temperature constitute a main problem for investigation. This work is still in progress. Some of the results obtained so far are given below.

4.1. Deuteron Irradiation at 200°K

Equation (1) shows that only those states with energies close to ξ are sensitive to changes of ζ in their effect on n–p. The variation in the rate of electron removal for the n-type samples, shown in Table 2, indicates that localized states are produced, having energies approximately $0 \cdot 1$ eV below the conduction band. From the variation of hole removal for the p-type samples we can deduce that energy levels within $0 \cdot 05$ eV of the valence band are introduced. Between $\zeta = E_C - 0 \cdot 17$ eV and $\zeta = E_V + 0 \cdot 1$ eV the effect of the irradiation on the carrier concentrations is comparatively small. It follows that bombardment introduces relatively few states with energy levels deep inside the energy gap at the start of irradiation, and that the number of states with energy levels near the valence band is nearly equal to the number of all donor states. If the energy levels near the valence band are due to acceptor states, then those near the conduction band must be donor states with the donors and acceptors equal in number.

4.2. Electron Irradiation

If one can assume that the number of defects is proportional to the flux of bombarding particles with no other complications, then instead of initial slopes for different samples we can use

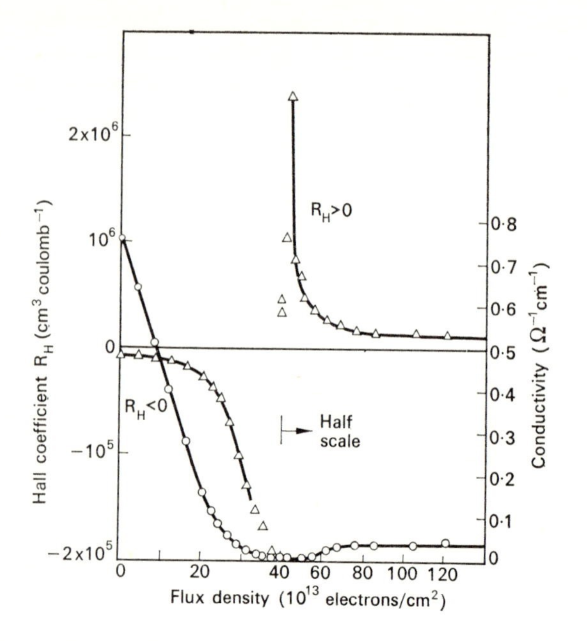

FIG. 3. Hall coefficients R and conductivity σ curves as a function of flux density for an initially n-type sample. Irradiation by 4·5 MeV electrons at 200°K. Δ—Hall coefficient R; ○—conductivity σ. Note the change in units of R (left-hand scale) on transition from negative to positive values. Also note the change in scale of flux density after conductivity minimum has been reached. (Measurements by E. Klontz and R. R. Pepper.)

continued irradiation of one sample. Figure 3 gives the conductivity and Hall coefficients of an originally n-type sample as a function of electron flux ϕ at 200°K. The conductivity decreased to a minimum with electron removal, then rose again as holes were added. The Hall coefficient, initially negative, increased first in magnitude as conduction electrons were reduced, changed its sign to positive as the conductivity reached a minimum. The Hall coefficient decreased in magnitude as holes were added.

From these curves the value of $(-\Delta n + \Delta p)/\phi$ and the value

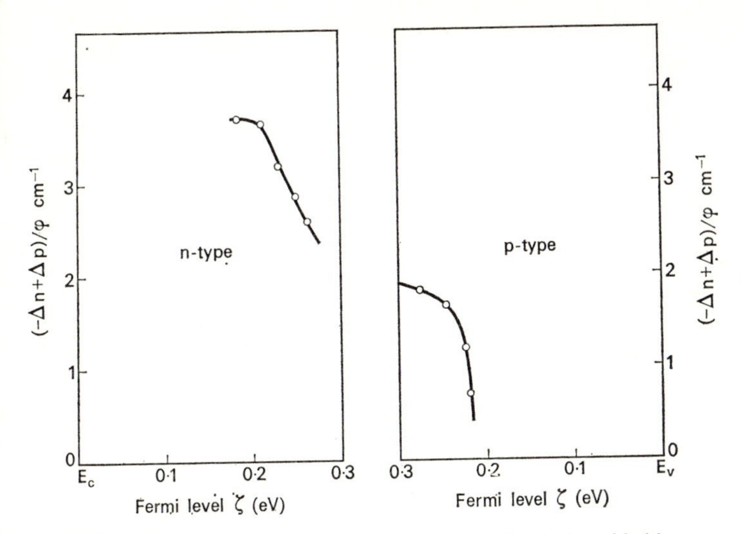

FIG. 4. Number of conduction electrons removed or holes added by a 4·5 MeV electron per unit length, as a function of the Fermi-level. Irradiation at 200°K. (Calculated by taking the total change of concentration (n cm^{-3}) at a time of irradiation t divided by the total flux cm^{-2} incident on the sample up to the time t.)

of ζ can be calculated for different stages of irradiation. The results of calculations made for a similar sample are plotted in Fig. 4, which shows that levels approximately 0·2 eV from the two edges of the energy gap were produced. Samples of high carrier concentrations, both n-type and p-type, have yet to be studied over a wider range of temperature in order to investigate the existence of levels nearer to the band edges as indicated by the deuteron irradiation. At 90°K such observations have been made with electron irradiation. However, at this low temperature difficulties of establishing equilibrium for the electron distribution arise and definite results cannot yet be reported, except that a drop of $(-\Delta n + \Delta p)/\phi$ is observed as ζ drops below approximately 0·1 eV from the conduction band, indicating that levels are also introduced in this energy range.

4.3. Irradiated Samples Warmed up to Room Temperature

For samples irradiated with deuterons at 90°K which have been warmed to room temperature, a typical (Hall coefficient, $1/T$) curve shows that the original *n*-type sample was converted to *p*-type and that the Fermi-level at 78°K was about 0·018 eV above the valence band. There must be then acceptor states very close to the valence band in excess of the original donor impurities plus donors due to the irradiations.

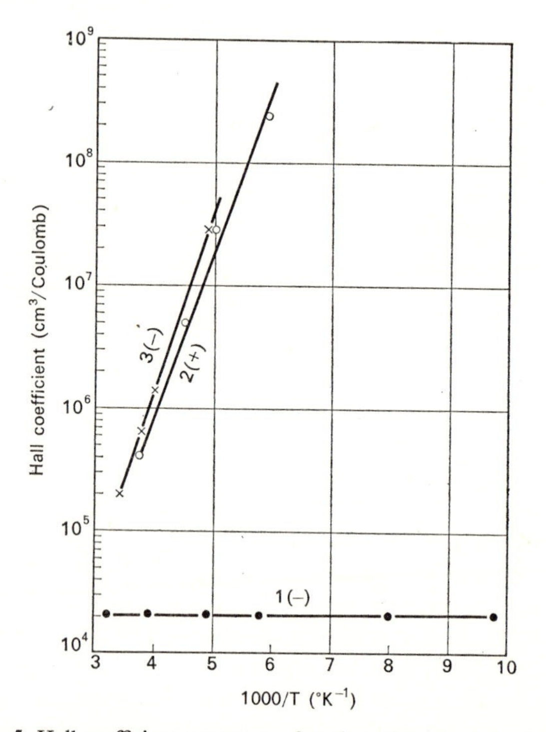

FIG. 5. Hall coefficient curves as a function of temperature for an initially *n*-type sample irradiated by 4·5 MeV electrons at 100°K. Curve 1, *n*-type, was taken before irradiation. Curve 2, *p*-type, was taken after the irradiated sample had been warmed to 0°C. Curve 3, *n*-type, was taken after the sample had warmed to room temperature. (E. Klontz and R. R. Pepper.)

Figure 5 gives the Hall coefficient curves for a sample irradiated by 4·5 MeV electrons at 100°K. The sample was originally *n*-type with Hall coefficient given by curve 1. Curve 2 was taken after the sample, following the irradiation, had been warmed up to 0°C; the sample was then *p*-type. The sample reverted to *n*-type after warming up to room temperature, the Hall coefficient being given by curve 3. The exponential rise of Hall coefficient with $1/T$ shown by curves 2 and 3 indicate the presence of energy levels approximately 0·22 eV from the valence band and from the conduction band. A similar curve has been reported for a sample irradiated by fast neutrons near room temperature, indicating a level approximately 0·1 eV above the valence band (Lark-Horovitz, 1951).

We can therefore state: (i) both donors and acceptors are produced by irradiation, (ii) levels very close to the valence band and approximately 0·1 eV below the conduction band were found as well as levels approximately 0·2 eV below the conduction band and approximately 0·1 and 0·2 eV above the valence band. The picture is not yet complete enough to warrant an attempt to correlate these levels with possible types of defects.

5. Defect Production as Function of Irradiation

It is possible that with prolonged irradiation the amount of defects produced may not increase linearly with flux. There may even be changes in the proportions of different types of defects with increasing irradiation. For electron irradiation at 90°K the following evidence indicates that such complications do not arise within the range of flux tested. Referring to equations (1), if all D's and A's are proportional to the flux ϕ, then $(-\Delta n + \Delta p)$ should increase linearly with ϕ, provided irradiation to the same carrier concentration has been carried out. (The same ζ level has been reached.) The results obtained by using *n*-type samples of different initial carrier concentrations are plotted in Fig. 6 which shows this linear relationship.

The situation seems to be different for deuteron irradiation. In

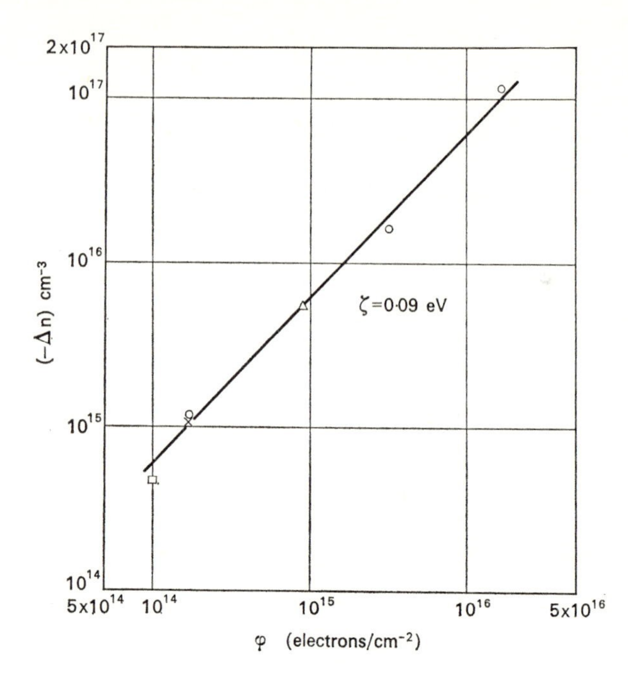

FIG. 6. Reduction of electron concentration: Δn as a function of flux density for different samples at a given Fermi-level, $\zeta = E_C - 0{\cdot}09$ eV obtained by irradiating different samples to the same final carrier concentration or ζ level corresponding to $E_C - \zeta = 0{\cdot}09$ eV. Irradiation by 4·5 MeV electrons at 90°K.

Fig. 7 the dots are obtained from continuous irradiation of an originally *n*-type sample and the crosses are obtained from the irradiation of a *p*-type sample. The curve shows that for a sample with ζ at 0·1 eV above the valence band such irradiations will have no effect on the carrier concentration. The circles give initial slopes $-\,dn/d\phi$ for the continuously irradiated *n*-type and two other *n*-type samples (see Table 2). If the various localized states increased linearly with flux the points for the two other samples should fall on the same curve. Furthermore, with simple "radiation annealing", i.e. if the defect production shows a saturation

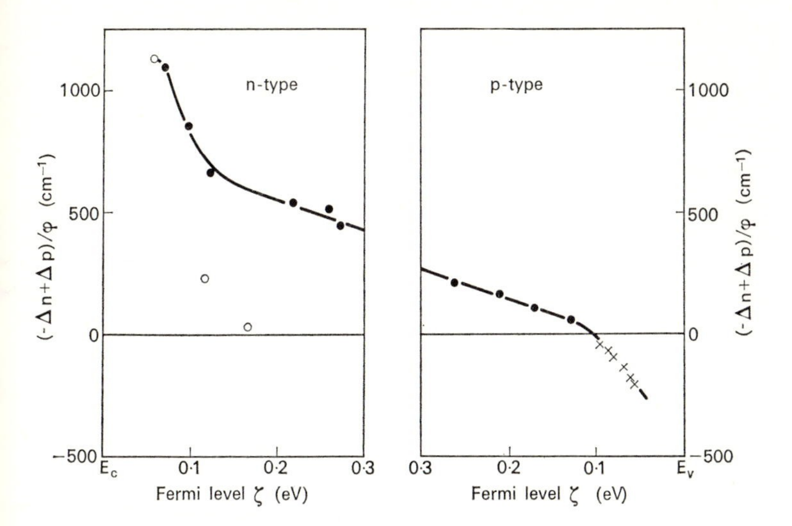

FIG. 7. Number of conduction electrons added or holes removed by a 9·6 MeV deuteron per unit length, as function of the Fermi-level. Irradiation at 200°K. Dots obtained on an originally *n*-type sample; crosses taken on a *p*-type sample. Circles give the data at the beginning of irradiation for three *n*-type samples. (Listed also in Table 2, column 2 and column 3.)

tendency but the various types of defects maintain the same ratio, the points representing the initial slopes should have been higher than the curve obtained from continuous bombardment, since the radiation annealing should have less effect at the start of the irradiation. The fact that they fall far below the curve indicates that relatively more acceptors were produced at longer irradiations.† As will be discussed in a later section, there is considerable

† Evidence to the same effect is given by some observations on a *p*-type sample. At the beginning of the irradiation at 200°K the conductivity decreased, showing that holes were removed. With increasing irradiation, however, the conductivity reached a minimum and increased steadily to approach a constant value. Such behaviour also indicates the effect of acceptors near the valence band at larger fluxes.

thermal annealing at this temperature, which should be in part responsible for these effects. However, irradiations carried out at 90°K also indicate that we do not have a simple increase of all types of defects proportional to the flux.

6. Non-equilibrium Distribution of Electrons in Localized States

Our discussion has been concerned so far with equilibrium distribution of electrons as given by equation (1). To avoid thermal annealing, experiments should be carried out at the lowest possible temperature. However, excitation due to the irradiation itself or photo effects may disturb this equilibrium, and if trapping mechanisms with long time constants exist, it is difficult to approach equilibrium conditions. Extra electrons in localized states may be referred to as trapped electrons and a deficiency may be referred to as trapped holes. The result of trapping is either a decrease of the number of carriers or an equal increase of carriers of the opposite sign, as stipulated by the condition of electrical neutrality. At low temperature there is a negligible concentration of minority carriers; an appreciable increase of these will be limited by the recombination with majority carriers. Thus trapping of majority carriers should result in a reduction of these carriers and decreased conductivity, whereas the trapping of the minority carriers will lead to an increase in majority carriers and increased conductivity. The following results indicate slow trapping effects in irradiated samples.

An *n*-type sample, which remained *n*-type after irradiation by 4·5 MeV electrons at liquid nitrogen temperature, was kept at this temperature. Upon illumination by white light the conductivity σ rose promptly, then kept increasing slowly and linearly with time, showing a tendency of saturation only after many hours. When the light was turned off there was a prompt drop of σ, equal in magnitude to the fast rise at the beginning of illumination. After this initial prompt decrease, σ was 38% higher than the original values. It then changed very slowly, and after

21 hours the difference was still 36%. The ratio of Hall coefficient to resistivity was the same before and after the illumination *but was 10% lower during the illumination*, which indicates that there was some hole conduction during illumination and that the fast part of the response was associated with the building-up and the decay of free holes. The slow changes must have been caused by the trapping of minority carriers, holes, and thus indicate slow rates of trapping and release.

An n-type sample, irradiated by 4·5 MeV electrons at 200°K temperature, was converted to p-type. There was a prompt increase in conductivity, when the sample was illuminated, which decayed in about a minute after the light was shut off. While the rise of conductivity remained fast, the decay time became longer as the sample temperature was reduced, 4 min at 185°K, 10 min at 170°K, and no appreciable recovery for several hours at 90°K. These results show that electrons, the minority carriers in p-type material, are trapped quickly but released slowly at low temperatures.

7. Thermal Annealing below Room Temperature

7.1. Electron Irradiation (4·5 MeV)

At 200°K the resistivity and Hall coefficient keep changing slowly after irradiation, in the direction toward the original values. From the values measured at any time t we can find from the curves taken during the irradiation what flux ϕ_e would be required to give these values. For a sample kept at 200°K for 70 hours, the corresponding ϕ_e was found to be 35% of the total flux used. The value of ϕ_e would therefore indicate the extent of annealing, if the pattern of defects has not changed.

The resistivity and Hall coefficient change with time after irradiation, even at 90°K. For samples converted from n-type to p-type, ρ and R increased by orders of magnitude in the course of a day. The change was in the direction back to n-type and might be taken as indication of simple annealing. After the sample was warmed up to about 130°K for several hours, the values of R and ρ at 90°K increased even higher.

However, if the sample is held a few hours at about 170°K, R and ρ dropped back close to the values at the end of the irradiation. This effect could be interpreted as a sign of changes in the defect pattern. However, in view of the discussions of trapping effects in the previous section, it is doubtful whether the changes in ρ and R, observed when the samples were kept at the lower temperatures, were really due to changes in the defect structure. This point of view is supported by the following observation. A sample irradiated at 200°K showed slow changes in ρ and R after having been cooled down to 90°K. The values dropped by an order of magnitude in several hours. When warmed up to 200°K the values of ρ and R were the same as before cooling. This process was repeated, giving the same results. Except for the observations at low temperatures with complications due to slow trapping processes, the changes in R and ρ for temperature above approximately 170°K were in the direction toward the original values before irradiation, and there is no indication of changes in the composition of defects.

7.2. Deuteron Irradiation

The behaviour during annealing of deuteron irradiated samples is quite different. An n-type sample converted to p-type by irradiation at 90°K was warmed up to successively higher temperatures and cooled down again to 77°K. Measurements of ρ and R were made as the sample was warmed up. The resistivity curves are given in Figs. 8(a) and 8(b). Beginning with curve 9, the following curves were taken after the sample was held at the highest temperatures sufficiently long for further changes to be slow. These curves were reproduced when the temperatures were varied between 77°K and the highest temperature of each curve. Figure 8(a) shows that the resistivity increased after successively higher temperatures had been reached. Some of these changes could be slow adjustments of non-equilibrium electron distribution.

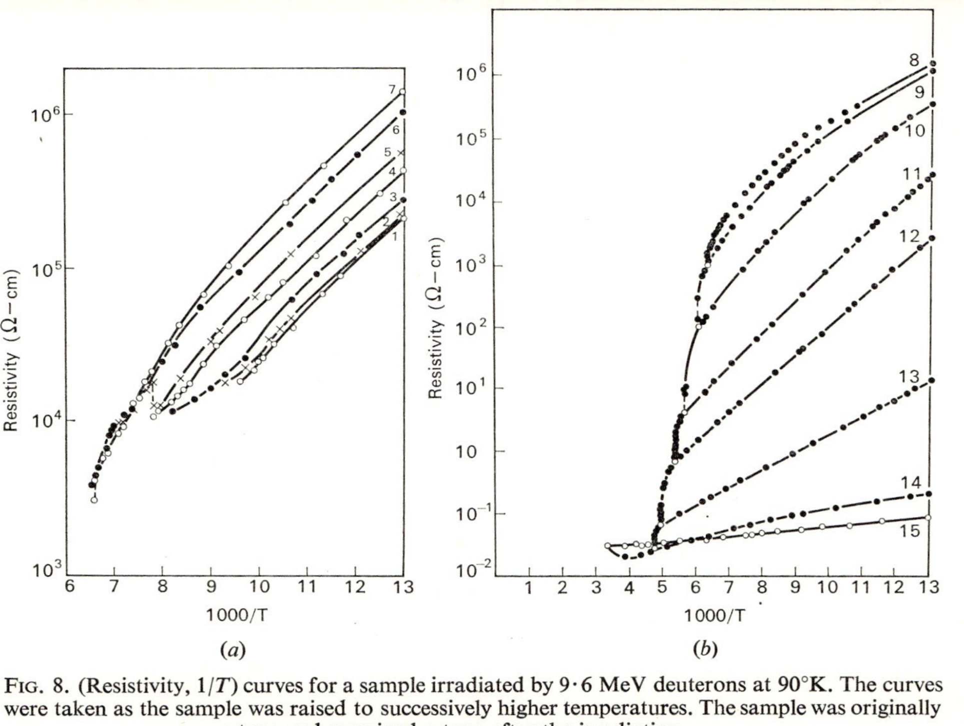

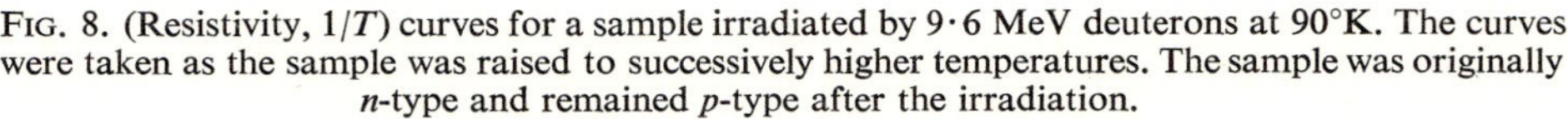
FIG. 8. (Resistivity, $1/T$) curves for a sample irradiated by 9·6 MeV deuterons at 90°K. The curves were taken as the sample was raised to successively higher temperatures. The sample was originally *n*-type and remained *p*-type after the irradiation.

Figure 8(*b*) shows the most interesting behaviour. These curves, which were reproducible, dropped lower with each higher temperature reached by the sample, beginning at about 160°K.† Similar behaviour was shown in the Hall coefficient curves. Thus thermal annealing increased the conductivity and hole concentration. The changes are in the direction opposite to returning to the original *n*-type conduction. After the room temperature is reached, the values of R and ρ when the sample is again cooled to 90°K were about six orders of magnitude lower than at the end of irradiation. Thermal annealing had carried the originally *n*-type sample to much higher *p*-type conduction than the irradiation at 90°K. In fact if we used a *p*-type sample of the same resistivity, irradiation at 90°K would have increased its resistivity. This behaviour clearly shows that temperature annealing changes markedly the pattern of defects produced by deuteron irradiation at the low temperature.

8. Effect of Irradiation on Mobility

We have considered so far only the effect of irradiation on the carrier concentrations, which plays the dominant role in determining the changes in electrical properties. Table 3 shows the effect on the mobility of conduction carriers. The values of mobility μ, calculated from the Hall coefficient and resistivity, are given before and after the irradiation. The two values in brackets were measured at the end of irradiation and before the sample was warmed up. N_c is the concentration of effective impurities in the samples. The type of conduction before irradiation is given in the fifth column. All samples were *p*-type after irradiation. Qualitatively there is correlation between the irradiation flux and the effect on the mobility. Sample 4, which received the least deuteron flux, shows the smallest effect in the mobility. Electron irradiation, which produces about 200 times smaller number of displacements (see Table 1) as compared with the

† Healing curves of this type were first observed by R. E. Davis in neutron irradiated silicon (see Johnson and Lark-Horovitz, 1948).

TABLE 3. THE EFFECT OF IRRADIATION ON CARRIER MOBILITY μ

	Sample	Flux (10^{16} cm^{-2})	N_c (10^{15} cm^{-3})	Type	Before irradiation μ (cm^2 V^{-1} sec^{-1})		After irradiation μ (cm^2 V^{-1} sec^{-1})		
					294°K	83°K	90°K	294°K	77°K
Deuteron	1	6·5	350	n	1080	844		549	508
9·6 MeV	2	3·5	980	p	550	439		415	284
200°K	3	3·3	4·8	p	1650	5980		344	490
	4	0·79	8·5	p	1620	8570		1000	3880
Deuteron 90°K	5	5	0·66	n	3600	27,000	(11)	270	257
Electron 90°K	6	2·1	0·27	p	1530	12,800	(4620)	1660	14,100

deuterons, shows also much smaller effect on the mobility. This is clear from the comparison of the two values of μ in brackets. Also after warming to room temperature, there was no significant effect of the electron irradiation on μ, whereas the samples irradiated by comparable fluxes of deuterons showed considerable reduction in mobility.

Comparison of the values of μ in brackets and the corresponding values at 77°K after warming the samples to room temperature shows the effect of thermal annealing. It is interesting to note the following point: sample 5 (discussed in the previous section) is converted from *n*-type to *p*-type by deuteron irradiation at 90°K; it became more *p*-type conducting by warming to room temperature. The change in μ produced by the warming shows clearly, however, that large amounts of defects were actually healed out. This can be understood if originally compensating defects of both types were present, but the donors are healed out, leaving excess acceptors behind.

References

Becker, M., Fan, H. Y., and Lark-Horovitz, K. (1952) *Phys. Rev.* **85,** 730.

Cleland, J., Lark-Horovitz, K., and Pigg, J. C. (1950) *Phys. Rev.* **78,** 814.

Dienes, G. J. (1953) *Ann. Rev. Nucl. Sci.* **2,** 187.

Fan, H. Y., and Becker, M. (1951) *Semi-conducting Materials*, Ed. H. K. Henisch (London: Butterworths Scientific Publications), pp. 132–148.

Fan, H. Y., Kaiser, W., Klontz, E. E., Lark-Horovitz, K., and Pepper, R. R. (1954) *Phys. Rev.* **95,** 1087.

Forster, J. H. (1952) Thesis, Purdue University.

Forster, J. H., Fan, H. Y., and Lark-Horovitz, K. (1952) *Phys. Rev.* **86,** 643; (1953) *ibid.* **91,** 229.

Heller, Z., and Tendam, D. (1951) *Phys. Rev.* **84,** 905.

James, H. M., and Lark-Horovitz, K. (1951) *Z. Phys. Chem.* **198,** 107.

Johnson, W. E., and Lark-Horovitz, K. (1948) *Radiation Damage Symposium, N.E.P.A.* 1178–ER–23.

Klontz, E. E., and Lark-Horovitz, K. (1951 a) U.S.A.E.C. Document T1D–5010; (1951 b) *Phys. Rev.* **83,** 763; (1952) *ibid.* **86,** 643.

Lark-Horovitz, K. (1951) *Semi-conducting Materials*, Ed. H. K. Henisch (London: Butterworths Scientific Publications), pp. 47–78.

Lark-Horovitz, K., Bleuler, E., Davis, R. E., and Tendam, D. (1948) *Phys. Rev.* **73,** 1256.

LARK-HOROVITZ, K., and BECKER, W. M. (1954) *Bull. Amer. Electrochem. Soc.*, p. 110.
MOTT, N. F. (1929) *Proc. Roy. Soc.* **A124,** 429.
PEPPER, R. R., KLONTZ, E. E., LARK-HOROVITZ, K., and MACKAY, J. (1954) *Phys. Rev.* **94,** 1410.
SEITZ, F. (1949) *Disc. Faraday Soc.* **5,** 271.

Index